# Guides to Professional English

**Series Editor:**
Adrian Wallwork
Pisa, Italy

For further volumes:
http://www.springer.com/series/13345

Adrian Wallwork

# Telephone and Helpdesk Skills

A Guide to Professional English

Adrian Wallwork
Pisa
Italy

ISBN 978-1-4939-0637-6          ISBN 978-1-4939-0638-3 (eBook)
DOI 10.1007/978-1-4939-0638-3
Springer New York Heidelberg Dordrecht London

Library of Congress Control Number: 2014939423

© Springer Science+Business Media New York 2014
This work is subject to copyright. All rights are reserved by the Publisher, whether the whole or part of the material is concerned, specifically the rights of translation, reprinting, reuse of illustrations, recitation, broadcasting, reproduction on microfilms or in any other physical way, and transmission or information storage and retrieval, electronic adaptation, computer software, or by similar or dissimilar methodology now known or hereafter developed. Exempted from this legal reservation are brief excerpts in connection with reviews or scholarly analysis or material supplied specifically for the purpose of being entered and executed on a computer system, for exclusive use by the purchaser of the work. Duplication of this publication or parts thereof is permitted only under the provisions of the Copyright Law of the Publisher's location, in its current version, and permission for use must always be obtained from Springer. Permissions for use may be obtained through RightsLink at the Copyright Clearance Center. Violations are liable to prosecution under the respective Copyright Law.
The use of general descriptive names, registered names, trademarks, service marks, etc. in this publication does not imply, even in the absence of a specific statement, that such names are exempt from the relevant protective laws and regulations and therefore free for general use.
While the advice and information in this book are believed to be true and accurate at the date of publication, neither the authors nor the editors nor the publisher can accept any legal responsibility for any errors or omissions that may be made. The publisher makes no warranty, express or implied, with respect to the material contained herein.

Springer is part of Springer Science+Business Media (www.springer.com)

# Introduction for the reader

## Who is this book for?

If you make telephone calls as part of your work, then this book is for you. Whether you work on reception, on a helpdesk, or simply telephone colleagues and clients, by applying the suggested guidelines, you will stand a much greater chance of making an effective telephone call.

The book is aimed at non-native English speakers, with an intermediate level and above.

I hope that other trainers like myself in Business English will also find the book a source of useful ideas to pass on to students.

This book is NOT for academics. Instead, read Parts III and IV of *English for Academic Correspondence and Socializing* (Springer Science), from which some of the subsections in this book are taken or adapted.


## What chapters should I read?

Receptionists / Switchboard Operators and Secretaries: Chapters 1-9, 14-18

Helpdesk / Technical: 1, 5, 8–12, 14–18

Sales and Marketing: 1–4, 6-9, 13–18

If you use the telephone for video conferences and audio conferences, read Chapter 16 of this book plus Chapter 6 (video conference calls) in the companion book *Meetings, Negotiations, and Socializing,* and Chapter 12 (audio conference calls) in *Presentations, Demos, and Training Sessions.*

Video conference calls - Chapter 6 in *Meetings, Negotiations and Socializing*

Audio conference calls - Chapter 12 in *Presentations, Demos and Training Sessions*


## How should I use the table of contents?

The table of contents lists each subsection contained within a chapter. You can use the titles of these subsections not only to find what you want but also as a summary for each chapter.

## Other books in this series

There are currently five other books in this *Guides to Professional English* series.

*CVs, Resumes, and LinkedIn*
http://www.springer.com/978-1-4939-0646-8/

*Email and Commercial Correspondence*
http://www.springer.com/978-1-4939-0634-5/

*User Guides, Manuals, and Technical Writing*
http://www.springer.com/978-1-4939-0640-6/

*Meetings, Negotiations, and Socializing*
http://www.springer.com/978-1-4939-0631-4/

*Presentations, Demos, and Training Sessions*
http://www.springer.com/978-1-4939-0643-7/

All the above books are intended for people working in industry rather than academia. The only exception is *CVs, Resumes, Cover Letters and LinkedIn*, which is aimed at both people in industry and academia.

There is also a parallel series of books covering similar skills for those in academia:

*English for Presentations at International Conferences*
http://www.springer.com/978-1-4419-6590-5/

*English for Writing Research Papers*
http://www.springer.com/978-1-4419-7921-6/

*English for Academic Correspondence and Socializing*
http://www.springer.com/978-1-4419-9400-4/

*English for Research: Usage, Style, and Grammar*
http://www.springer.com/978-1-4614-1592-3/

# INTRODUCTION FOR THE TEACHER

## Teaching Business English

I had two main targets when writing this book:

- non-native speakers (business, sales technical)
- Business English teachers and trainers

My teaching career initially started in general English but I soon moved into Business English, which I found was much more focused and where I could quickly see real results. The strategies I teach are almost totally language-independent, and many of my 'students' follow my guidelines when making phone calls (writing emails, presenting etc) in their own language. I am sure you will have found the same in your lessons too.

Typically, my lessons cover how to:

1. make phone calls
2. write emails
3. make presentations and demos
4. participate in meetings
5. socialize

This book is a personal collection of ideas picked up over the last 25 years. It is not intended as a course book; there are plenty of these already. It is more like a reference manual.

I also teach academics how to interact with colleagues around the world. In fact, a couple of the chapters in this book are based on chapters from *English for Academic Correspondence and Socializing* (Springer).

# How to teach telephone calls

I suggest you adopt the following strategy.

In your first lesson on telephoning, have a general discussion on:

- how much time your students spend on the telephone
- what their main fears are and how they manage to overcome them
- what typical calls they have to make / receive
- what useful phrases (Chapter 16) they know, and whether they have made their own personal collections of such phrases
- what preparation, if any, they make before making / receiving phone calls

Use ideas from Chapter 1 to help you guide the last part of the discussion.

In my experience the basic problems are:

1. lack of preparation
2. not knowing the right phrases to use
3. inability to understand the caller
4. nerves

Nerves are caused by points 1 - 3.

To resolve POINT 1, you need to help your students learn how to simulate a call in advance (see Chapter 1.5). The best way to do this is to choose two students who do a similar job e.g. they both work on reception, or both work on the helpdesk. Get them to think of the typical calls they make (in English and in their own language). Then ask them to simulate the call in their own language. Make notes of the structure of the call, while the two students are talking. Then gradually build up the same dialog in English (either on your laptop or the whiteboard) - refer to your notes to make sure that you don't forget to include something. On the basis of this dialog, they can then practise what to say. It makes sense at this stage of your telephoning course to have very small groups of people all of whom do they same job, otherwise some of the dialogs will be totally irrelevant for other students.

Resolving POINT 1 goes some way to resolving POINT 2 at the same time. While building up the dialogs, introduce useful phrases from Chapter 16. Encourage students to learn the phrases that they find the easiest to remember and easiest to say. But they need to be aware of the other phrases in case their interlocutor uses them.

POINT 3 - inability to understand the caller - is certainly the biggest problem. Firstly, students need to understand that a conversation is two way. If they don't understand something the responsibility is not entirely their own, but also rests with their interlocutor. It is worth stressing this concept many times to ensure that your students do not feel stupid or humiliated when they don't understand the caller. Instead, they should remain calm and adopt a series of strategies (Chapter 11) to help them try and understand the caller better. Chapter 12 (on pronunciation and word stress) and Chapter 13 (using the web to improve listening skills), will also help them understand better. You can also use the listening exercises contained in the many telephone skills books produced by ELT publishers such as CUP, OUP, Macmillan and Longman.

If you work in-house, then you can help your students massively if you listen in on their calls. When you debrief them after the call, you can then help them to improve their technique. So get involved with the company / companies where you teach. You will find it much more satisfying!

# Contents

**1 PREPARING FOR CALLS** ............................................................. 1
   1.1  Before calling, decide whether another form of communication might be more suitable ................................... 1
   1.2  Use email as a preliminary information exchange before the call ........................................................................ 2
   1.3  Accept that you might feel nervous: prepare and practice ......... 2
   1.4  Simulating, recording and transcribing telephone calls ............. 3
   1.5  Making a good first impression ................................................. 4
   1.6  If you receive a call, don't be afraid to tell the caller that this is a bad time to talk ...................................................... 5
   Preparing for the call: Summary ......................................................... 6

**2 MAKING A CALL** ....................................................................... 7
   2.1  Give your name and the name of the person you want to talk to ..................................................................... 7
   2.2  When you have been connected, explain who you are and the reason for your call ..................................................... 8
   2.3  Take notes, summarize and follow up with an email ................. 9
   2.4  What to do if your English is high level but your interlocutor's is low level ........................................................... 10
   Making a call: Summary .................................................................... 11

**3 LEAVING A MESSAGE WITH THE SWITCHBOARD OPERATOR** ................................................................................ 13
   3.1  Learn the structure and typical phrases of a telephone message ....................................................................................... 13
   3.2  Speak clearly and slowly ............................................................ 15
   3.3  Make the call as interactive as possible .................................... 15
   3.4  Spell names out clearly using the International Alphabet or equivalent ............................................................. 16
   3.5  Practice spelling out addresses ................................................. 18
   3.6  When spelling out telephone numbers, read each digit individually .................................................................................. 20

3.7 Consider sending a fax, rather than an email, confirming what has been said ................................................. 20
Leaving a telephone message: Summary ........................................... 21

## 4 VOICEMAIL AND ANSWERING MACHINES ................................. 23
4.1 Use an appropriate voicemail ....................................... 23
4.2 Leaving a message on someone's answering machine ............ 25
Voicemail and answering machines: Summary .................................. 26

## 5 RECEIVING CALLS: WORKING ON RECEPTION / SWITCHBOARD ............................................................................ 27
5.1 Initial salutations ........................................................ 27
5.2 Transferring the call for a client .................................... 28
5.3 Transferring the call for a colleague: informal version .............. 30
5.4 Transferring the call for a colleague: more formal version ......... 31
5.5 Creating a friendly relationship with colleagues ....................... 31
5.6 Choosing the easiest phrase to say ................................. 32
5.7 Use of *will* and present perfect ................................... 33
5.8 Being proactive and helpful ......................................... 34
5.9 Adopting a friendly tone ............................................. 35
5.10 Taking a message ...................................................... 36
5.11 Dealing with wrong numbers ........................................ 37
Receiving a call: Summary ............................................................ 38

## 6 FINDING OUT ABOUT ANOTHER COMPANY, GIVING INFORMATION ABOUT YOUR COMPANY ...................................... 39
6.1 Responding to a caller who wants information about your company ........................................................... 39
6.2 Calling a company to find out information about that company .......................................................... 41
6.3 Calling someone in a company to make a cold sale ................. 41
Finding out about another company, giving info about your own company: Summary ............................................................... 43

## 7 CHASING ...................................................................... 45
7.1 Chasing a payment ..................................................... 45
7.2 Chasing an order ....................................................... 49
7.3 Chasing a document, report etc. ................................... 50
Chasing: Summary ..................................................................... 52

## 8 DEALING WITH DIFFICULT CALLERS AND UNHELPFUL STAFF ................................................................ 53
8.1 Dealing with people who are trying to sell you a product / service that your company is not interested in ............ 53
8.2 Dealing with people who are waiting for a response from someone within the company but have had no reply ........ 54

|  |  |  |
|---|---|---|
| 8.3 | Switchboard: dealing with a client who wants to register a complaint | 56 |
| 8.4 | Person responsible: dealing with a client who has received poor service | 57 |
| 8.5 | Dealing with rude callers | 58 |
| 8.6 | Dealing with unhelpful staff when you are the caller | 59 |
| | Dealing with difficult callers and unhelpful staff: Summary | 60 |

## 9 IMPROVING YOUR TELEPHONE MANNER ........... 61

| | | |
|---|---|---|
| 9.1 | Avoid being too direct | 61 |
| 9.2 | Help the person that you want to speak to | 62 |
| 9.3 | Speak slowly and clearly | 62 |
| 9.4 | Don't be afraid to interrupt and make frequent summaries of what you think you have understood | 63 |
| 9.5 | Compensate for lack of body language | 63 |
| 9.6 | Learning to sound authoritative and competent | 64 |
| 9.7 | Evaluate your performance | 65 |
| | Improving your telephone manner: Summary | 66 |

## 10 WORKING ON A HELPDESK: KEY ISSUES ........... 67

| | | |
|---|---|---|
| 10.1 | Do not panic. Listen to the full explanation before reacting | 67 |
| 10.2 | Admit that you have not understood | 69 |
| 10.3 | Improve your pronunciation | 69 |
| 10.4 | Ask the caller to speak more slowly | 69 |
| 10.5 | Check whether the caller has a single problem or a multiple problem | 70 |
| 10.6 | Prepare possible customer questions and solutions to these questions | 70 |
| | Key helpdesk issues: Summary | 71 |

## 11 HELPDESK: DEALING WITH CUSTOMERS ........... 73

| | | |
|---|---|---|
| 11.1 | Dealing with a customer's problem: a ten-step solution | 73 |
| 11.2 | What to say while the customer is explaining the problem | 75 |
| 11.3 | Using questions to identify the problem | 76 |
| 11.4 | Interrupting and repeating back what the customer tells you | 77 |
| 11.5 | Suggesting possible causes and solutions: expressing certainty through adverbs and modal verbs | 79 |
| 11.6 | Giving instructions to the customer | 80 |
| 11.7 | Telling the customer what you need from them and what the next step will be | 81 |
| 11.8 | Showing the customer that you care | 82 |
| 11.9 | Follow up with an email | 83 |
| | Helpdesk – dealing with customers: Summary | 84 |

## 12 CALLING A HELPDESK .................................................................. 85
- 12.1 Facilitating a smooth service from the helpdesk operator ..... 85
- 12.2 How to interact with a helpdesk operator who has very poor English ................................................................ 87
- Calling a helpdesk: Summary .......................................................... 88

## 13 PARTICIPATING IN AUDIO AND VIDEO CONFERENCE CALLS ................................................................. 89
- 13.1 Audio conference calls ................................................................ 90
- 13.2 Preparing for a conference call .................................................. 91
- 13.3 Introducing yourself ................................................................... 92
- 13.4 Dealing with technical and documentation problems ............ 93
- 13.5 Checking for clarifications during the call ............................... 94
- 13.6 Ending the call ........................................................................... 95
- 13.7 Skype calls .................................................................................. 96
- Participating in audio and video conferences: Summary .................. 97

## 14 WHAT TO DO AND SAY IF YOU DON'T UNDERSTAND .............. 99
- 14.1 Foreign language skills of native English speakers ................ 99
- 14.2 Ignoring words and expressions that you don't understand ..................................................................... 101
- 14.3 Don't say 'repeat please' ......................................................... 103
- 14.4 Choose the quickest and easiest way to indicate exactly what you don't understand ......................................... 105
- 14.5 More examples of asking for clarification and making comments ................................................................................. 106
- 14.6 Distinguish between similar sounding words ........................ 108
- 14.7 Use instant messaging systems ............................................. 109
- 14.8 If you really can't understand, learn a way to close the call ............................................................................ 109
- 14.9 Reasons why understanding a native English speaker can be difficult ............................................................ 110
- Improving your understanding of what the caller is saying: Summary ........................................................................................ 114

## 15 USING THE WEB AND TV TO IMPROVE YOUR LISTENING SKILLS ....................................................................... 115
- 15.1 Set yourself a realistic objective ............................................. 115
- 15.2 The news ................................................................................... 116
- 15.3 YouTube .................................................................................... 116
- 15.4 Dragon's Den ............................................................................ 116
- 15.5 TV series ................................................................................... 117
- 15.6 TED ............................................................................................ 118
- 15.7 Movies ...................................................................................... 119
- 15.8 Subtitles ................................................................................... 120
- 15.9 Songs ........................................................................................ 120

|   |   |   |
|---|---|---|
| 15.10 | Audio books and podcasts | 120 |
| 15.11 | Other websites worth checking out | 120 |
|   | Using the web and TV to improve your listening skills: Summary | 121 |

## 16 PRONUNCIATION: WORD AND SENTENCE STRESS ... 123

|   |   |   |
|---|---|---|
| 16.1 | Investigate free software that will help you to improve your pronunciation | 123 |
| 16.2 | Two syllables: general rules | 125 |
| 16.3 | Two syllables: same word (noun on first, verb on second) | 126 |
| 16.4 | Compound nouns | 126 |
| 16.5 | Three syllables | 127 |
| 16.6 | Multi-syllable words | 128 |
| 16.7 | Acronyms | 128 |
| 16.8 | Sentence stress | 129 |

## 17 EXAMPLE TELEPHONE DIALOGS ... 131

|   |   |   |
|---|---|---|
| 17.1 | Switchboard operator: trying to connect someone | 131 |
| 17.2 | Switchboard: Taking down someone's name and number | 132 |
| 17.3 | Switchboard: checking understanding | 133 |
| 17.4 | Switchboard operator: chit chat | 133 |
| 17.5 | Switchboard: dealing with an employee who rings in sick | 134 |
| 17.6 | Switchboard: giving out a phone number | 135 |
| 17.7 | Switchboard: dealing with a caller whose request you cannot fulfill | 136 |
| 17.8 | Leaving a message with the switchboard | 136 |
| 17.9 | Arranging a meeting | 137 |
| 17.10 | Changing the time of a meeting | 137 |
| 17.11 | Sales division: Dealing with a customer inquiry | 138 |
| 17.12 | Making an enquiry about a company | 140 |
| 17.13 | Outlining / Solving technical problems 1 | 141 |
| 17.14 | Outlining / Solving technical problems 2 | 141 |
| 17.15 | Outlining / Solving technical problems 3 | 142 |
| 17.16 | Extracts from a conference call | 143 |

## 18 USEFUL PHRASES ... 145

|   |   |   |
|---|---|---|
| 18.1 | Switchboard: saying / establishing who is calling | 145 |
| 18.2 | Calling: saying who you are and who you want to speak to | 148 |
| 18.3 | Calling: when person desired is not available | 149 |
| 18.4 | Initiating the call with the desired person | 151 |
| 18.5 | Leaving a message | 151 |
| 18.6 | Taking a message | 153 |
| 18.7 | Problems with understanding | 154 |
| 18.8 | Checking and clarifying | 155 |
| 18.9 | Calling someone you already know: giving background to your call / updating | 156 |

- 18.10 Calling someone back .......................................................... 159
- 18.11 Requests / enquiries ............................................................ 159
- 18.12 Cold calling (calling a company for the first time) .................. 161
- 18.13 Making a complaint, registering a problem, calling a helpdesk.... 162
- 18.14 Helpdesk: finding out about the problem ............................. 163
- 18.15 Helpdesk: dealing with a problem ........................................ 164
- 18.16 Helpdesk: checking that you have both understood each other .................................................................................. 165
- 18.17 Helpdesk: summarizing the problem, outlining a solution ..... 166
- 18.18 Being a good listener: reassuring and empathizing .............. 167
- 18.19 Apologizing ......................................................................... 168
- 18.20 Thanking ............................................................................. 169
- 18.21 Arranging a meeting for yourself .......................................... 169
- 18.22 Arranging a meeting for a colleague .................................... 171
- 18.23 Hotel reservations ............................................................... 172
- 18.24 Saying goodbye .................................................................. 173

**THE AUTHOR** ............................................................................. 175

**Index** ........................................................................................... 177

# 1  PREPARING FOR CALLS

## 1.1  Before calling, decide whether another form of communication might be more suitable

First, consider whether your phone call is really necessary. Decide whether it wouldn't be simpler, at least for your counterpart, for you simply to send an email or fax.

It is generally a good idea to speak to someone directly on the phone rather than send an email:

- if you want to establish a good relationship
- to solve any misunderstandings that have already arisen via email

You are unlikely to have the person's full attention if you call them on their mobile: you may well be disturbing them in the middle of something else. It is generally a good idea to ask:

> *Is this a good time or are you in the middle of something?*
>
> *Am I interrupting something?*

If they then say *Well, actually I am with someone at the moment. But go ahead, what can I do for you?* it is probably best to call back later and say:

> *Sorry, I have obviously got you at an inconvenient time. What time do you think I could call you back?*

## 1.2 Use email as a preliminary information exchange before the call

The more both parties are prepared for a telephone call, the more likely the call will be successful. If you have a call that will require a complex discussion, send each other a list of points that you wish to discuss. This will enable you to:

- think about what you need to say and how to say it
- think about what useful phrases in English you may need
- tick the items from the list as you discuss them, and make notes next to each item

You could suggest such an email exchange by writing:

> *Before we make our call, I thought it might be useful to send you this list of items that I would like to discuss. If you have any additions I would be glad to receive them. Then it would be great if you could give me a few hours to look through them. Thank you.*

## 1.3 Accept that you might feel nervous: prepare and practice

It is perfectly normal to feel anxious about making or receiving a phone call. It may help you to know that your interlocutor too may be feeling nervous!

Do some breathing exercises to calm your nerves before you make / receive the call.

Reassure yourself that if you don't understand anything, you can always clarify issues via email.

You will certainly feel less nervous about making a call if you prepare some notes about what you want to say, and then make sure you know how to say everything in English.

Think about what the other person might ask you, and prepare answers to such questions. If you do so, you are more likely to be able to understand the question when it is asked.

It is important not only to practise what you want to say, but to prepare for what you might hear, for example the typical phrases that a switchboard operator or secretary might say.

> *Could you tell me what it is in connection with?*
>
> *Has she spoken to you before?*
>
> *I'll just check for you. Could you hold for a moment?*

## 1.4 Simulating, recording and transcribing telephone calls

It is very useful to simulate and record (i.e. with audio) possible future telephone conversations. You can do this with a colleague or with your English teacher, either in your own language or in English. You can then transcribe / translate what you said, and make improvements to it. Possible improvements include:

- giving more precise and concise explanations
- giving more details in case these are required
- asking better formulated questions
- perfecting the English grammar, syntax and vocabulary
- choosing words that are easy for you to say
- keeping everything as simple as possible

Depending on your role in the conversation, i.e. whether you will primarily be asking or answering questions, you will either need to write questions or think of answers to questions.

If you have prepared the questions in advance, you:

- are more likely to hear them and understand the questions when they are asked
- will seem very professional because you will have clear concise and detailed answers to the questions
- will be more fluent and confident when you speak

## 1.5 Making a good first impression

We tend to form first impressions very quickly. In just a few seconds we decide whether we think someone is professional / competent or not, whether we feel we will be listened to or not, and whether we like a person or not. Even in a non-video call where we cannot see the other person, we still create a mental picture of them. And of course, the other person forms a picture of us too.

This impression extends from us to our organization. When you make a call, you are reflecting the organization you work for.

However, on the telephone we lose some of the clues to evaluating someone that we would have when we talk to someone face to face. This means that we cannot see the expression on our interlocutor's face, or their reactions to what we say. So we may reach the wrong conclusions in the same way as we often misinterpret the tone of an email.

But not seeing our interlocutor has advantages too.

Our interlocutor cannot see if you are looking really nervous. They cannot see if you are desperately trying to find the answer to their question by scrolling pages on your computer.

## 1.6 If you receive a call, don't be afraid to tell the caller that this is a bad time to talk

If you are feeling very nervous or not prepared for a phone call which you judge to be important, then consider asking the caller to call back. There are many excuses you can make:

*Sorry, I am actually in a meeting at the moment, do you think you could call back in an hour?*

*Sorry, but actually this is not a good time for me. Can I call you back this afternoon?*

*Sorry, I am just about to go into a meeting. Would you mind emailing me or calling me tomorrow morning?*

If you are feeling very nervous about talking on the phone in English and feel that the caller could probably provide the information via email, then suggest that he / she sends you a mail.

*Sorry but I can't take your call at the moment. Could you possibly email me instead? That way I could certainly get back to you quickly.*

As highlighted by the above phrase, you can motivate the caller to send you an email rather than speak on the phone, by informing them that you will actually be able to deal with them faster by email.

Clearly, this strategy can only be used when you know the call is for you personally, i.e. you are not someone working on reception.

## Preparing for the call: Summary

- Consider using email as an alternative to a phone call.
- Send a preliminary email before the call, outlining what you want to talk about and asking the other person to do the same. On this basis, prepare your questions and answers in English.
- For important calls, simulate with a colleague in advance.
- Remember that first impressions count
- Learn how to sound authoritative
- Get a colleague to assess your performance
- Consider delaying an important phone conversation by asking the caller to call back later.

# 2 MAKING A CALL

## 2.1 Give your name and the name of the person you want to talk to

Announce to the switchboard operator who you are slowly and clearly. Example:

> This is Riccardo Rizzi, that's R-I-Z-Z-I, calling from Ferrari, in Modena, Italy; can I speak to Andrea Caroli please.

It's always a good idea to give both the first name and surname of the person you want to speak to. First, your listener will have a greater chance of understanding who it is you want and secondly if you had asked for Ms Caroli or Mrs Caroli, you would have totally confused the operator because Andrea (in Italy at least) is a man's name, not a woman's name.

For spelling out names see 3.4.

## 2.2 When you have been connected, explain who you are and the reason for your call

When you have been connected to the right person, if this is the first time you have spoken, the normal practise is to:

1. announce who you are
2. who you work for
3. say how you got your interlocutor's name
4. explain why you are calling

The third point is not strictly necessary, but may motivate your interlocutor to listen to what you have to say.

You can easily prepare what you are going to say in advance. Here is a possible model:

> *This is Yohannes Gedamu calling from ABC. Your name was given to me by Anh Nguyet Tran [who thought you might be interested in...]. The reason I am calling is ...*

Before you announce who you are, you might also want to check that you are speaking to the right person:

> *Am I speaking to Dubravka Cukrov?*
>
> *Is that the manager of the business development team?*

Below is a longer dialogue that shows the possible evolution of a phone call between two people who work for the same team in a company but have never met or spoken to each other before. The caller is in *italics*.

> Suzanne Graves.
>
> *Good morning Ms Graves / Suzanne. This is Manfred Raspapovic from the Munich Office.*
>
> Oh, good morning Mr Raspapovic.
>
> *Oh please call me Manfred. I don't think we've actually spoken before.*
>
> No, I don't think we have. I met Wolfgang when he was over here last year, but I think you and I have just emailed. Good to talk to you at last.
>
> *So the reason I'm calling you is because Wolfgang and I have been working on a project that we wanted to discuss with you... [details of project etc]... Well, look, I'd better be going, I don't want to keep you too long.*
>
> Don't worry, that's fine.
>
> *Please could you send my regards to the team.*
>
> Certainly, and it was nice to talk to you.
>
> *And you. Goodbye.*
>
> Bye.

## 2.3 Take notes, summarize and follow up with an email

If you take notes during the call, you can occasionally repeat back what the other person has said so that you can check your understanding. Obviously notes will also help you to remember what was said. This will be useful if / when you send your interlocutor an email summarizing the call.

At the end of the call, in order to check that you have not missed anything, make a mini summary of what has been said. This gives your interlocutor an opportunity to clarify any points. You can say:

> *Can I just check that I have got everything? So we have decided to delay the second stage of the project. We are going to call the client to inform them why there will be this delay. We have set ourselves a deadline for the first stage for the end of this month. Does that all sound right?*

You and your interlocutor may remember little more than 10 % of what was said during your call. Even if you think the phone call has gone well and that you have understood everything, it is always good practice to send your interlocutor an email summarizing the main points. You can use the same words as in your summary at the end of your call. This allows the interlocutor to check his / her understanding of the call as well. In addition, you can ask any questions or clarify points that you forgot to make during the call itself.

## 2.4 What to do if your English is high level but your interlocutor's is low level

Give your interlocutor (i.e. the person you are speaking to) time to tune into your voice before you start asking for or giving important information. Remember they not only have to switch their brains onto what you're saying but they also have to change language.

Put the person at ease at the start of the call:

*Tell me to stop if I'm speaking too quickly or if you don't understand something.*

*Please ring me back if you realize afterwards that you haven't understood something.*

Speak slowly. This doesn't mean just putting pauses between each word, it also means clearly articulating each individual word too as naturally as possible. Be careful not to cut out words in your sentences in an attempt to make it simpler.

Don't assume that just because they speak well they understand what you are saying.

Never try to rush the phone call—give the other person time to digest information.

Offer to email a short summary of the main points after the phone call.

## Making a call: Summary

- At the beginning of the call, say: who you are, who you work for, why you are calling, and who you want to speak to.
- Check you are speaking to the right person.
- Take notes. Use these notes to make summaries during and at the end of the call, and in the follow-up email.
- Email the interlocutor with a summary.

# 3 LEAVING A MESSAGE WITH THE SWITCHBOARD OPERATOR

## 3.1 Learn the structure and typical phrases of a telephone message

The following example is designed to show you the language used in a typical telephone conversation, where one party leaves a message.

ABC. Good afternoon can I help you?

*Yes, this is Irmin Schweiz from XYZ. Could I speak to Wei Li please?*

Sorry I didn't catch your name, could you speak up a bit please, the line's bad.

*Yes, it's Irmin Schweiz.*

And where did you say you are ringing from?

*XYZ.*

OK, I'll try and connect you. ... Sorry, the line's busy. Do you want to leave a message?

*Could you tell her that Irmin Schweiz called, and that the meeting's been postponed till next Tuesday.*

That's Tuesday the seventh right?

*Right. But if she needs to speak to me he can get me on 0049 that's the code for Germany, then 89 656 2343. Extension fifteen.*

That's one five right?

*That's it.*

Can I read that back to you to make sure I've got everything?

*Sure.*

Irmin Schweiz, that's S-C-H-W-I-E-Z from the ...

*No, it's E-I not I-E.*

## 3.1 Learn the structure and typical phrases of a telephone message (cont.)

OK, from XYZ. The meeting's been postponed till Tuesday the seventh, and she can reach you on 0049 89 656 2334

*Sorry that should be four three, not three four.*

OK 2343, extension 16, one five.

*That's it. Thanks very much. Bye.*

Goodbye.

## 3.2 Speak clearly and slowly

If you need to leave a message, speak very slowly and clearly. Repeat each bit of information at least twice, particularly any numbers. Even the most proficient English speakers often take down the wrong numbers. If you need to mention days of the week, especially Tuesday and Thursday which are very easily confused, then always say the day with the date: 'Thursday the 16th', so that there's a greater chance of being understood.

## 3.3 Make the call as interactive as possible

If you want to ensure that your interlocutor carries out your requests, it helps if you encourage them to be active rather than passive. You can ask your interlocutor to:

- confirm what you have said
- to read back any email or website addresses to you
- to repeat the spelling of names

These techniques force the interlocutor to pay more attention. Conversely, if you are the receiver of the message then you can follow exactly the same techniques. You can say:

> *Can I just confirm what you have said. So, the meeting has been moved to ...*
>
> *I'll just read back your website address. www dot u-n-i-p-i slash ....*
>
> *So it's Anna Southern, that's S-O-U ...*

## 3.4 Spell names out clearly using the International Alphabet or equivalent

When spelling a word, make sure you differentiate clearly between easily confused letters such as B and P, and D and T, and N and M. There is an International Alphabet (see first column below), but few people are familiar with it. So you might find it easier to use the names of countries (second column), which you may also find easier to pronounce.

Note: Where the name of the country is not commonly known or could easily be spelt very differently in another language, I have suggested another word.

Always repeat the spelling and always do so slowly.

| | | |
|---|---|---|
| A | Alpha | Argentina |
| B | Bravo | Brazil |
| C | Charlie | Congo |
| D | Delta | Denmark |
| E | Echo | Ecuador |
| F | Foxtrot | France |
| G | Golf | Germany |
| H | Hotel | Holland (hotel) |
| I | India | India |
| J | Juliet | Japan |
| K | Kilo | Kenya |
| L | Lima | Lebanon (lemon) |
| M | Mike | Mexico |
| N | November | Norway |
| O | Oscar | Oman (orange) |
| P | Papa | Panama |
| Q | Quebec | Qatar (quick) |
| R | Romeo | Russia |
| S | Sierra | Spain |
| T | Tango | Turkey |

## 3.4 Spell names out clearly using the International Alphabet or equivalent (cont.)

| | | |
|---|---|---|
| U | Uniform | Uganda |
| V | Victor | Venezuela |
| W | Whisky | Wales (Washington) |
| X | X-ray | |
| Y | Yankee | Yemen (yellow) |
| Z | Zulu | Zambia |

When you spell out a name, you can say:

> My name is Schmidt. That's S as in Spain, C as in Congo etc

Alternatively:

> My name is Schmidt. That's Spain, Congo etc

Be especially careful when spelling out vowels, whose sounds tend to be very different from country to country.

Note how the following letters have the same final sound. In the second column there are common words with the same sound.

| | |
|---|---|
| b, c, d, e, g, p, v, z | be, see, we |
| j, k | way, say |
| i, y | I, my, why |
| q, u, w | you, two |

The letters have the same initial sound, which is the same as the vowel sound (in italics) in the example words.

| | |
|---|---|
| a, h | pl*ay*, s*ay*, w*ay* |
| f, l, m, n, x | b*e*d, g*e*t, w*e*ll |
| o | g*o*, n*o*, sh*ow* |

## 3.5 Practice spelling out addresses

Being able to spell out your email address without confusing your listener is a key skill when participating in a telephone call. This is because if communication between you and your interlocutor is difficult or impossible due to language difficulties, the easiest solution is to continue the communication via email. To be able to continue via email, at least one of you needs to give their email address. In reality you increase your chances of continuing the communication if you both give your addresses. Here is an example dialogue of someone giving their email address to a receptionist (in italics).

My address is anna_southern at virgilio dot it. That's anna A-N-N-A underscore ...

*Sorry, what is after ANNA?*

Underscore.

*Underscore?*

The little line between two words.

*OK.*

So, underscore then Southern. That's S as in Spain, O, U, T as in Turkey, H, E, R, N.

*Is that N as in Norway?*

Yes, that's right. Then at virgilio dot it. That's V, I, R, G, I, L, I, O dot I, T.

*OK. I'll just repeat that. 'anna' that's with two Ns right?*

Yes, A double N A.

*Then underscore S, O, U, T, H, E, R, N. So annasouthern, that's all one word, right?*

Yes, that's right-

*Then at virgilio, that's V, I ...*

If your address is rather complicated (e.g. with an underscore, slashes, or very long), it is not a bad idea to have a personal email address that is short and which is simple to say, which you can use for emergencies!

Here is an example dialogue of someone giving their website address:

So it's www englishconferences forward slash R1256 dot pdf.

*Sorry does englishconferences have a dot between the two words?*

No it's all one word. Then forward slash. The letter R. Then the numbers 1, 2, 5, 6 as digits not as words. Then dot pdf. Have you got that?

*I'll just read it back to you. So, that's forward slash ...*

## 3.5 Practice spelling out addresses (cont.)

Below is an example of how to dictate the following traditional postal address over the phone: Adrian Wallwork, Via Murolavoro 17, 56127 Pisa, Italy

> Adrian Wallwork. That's a-d-r-i-a-n new word w-a-l-l-w-o-r-k. New line. Via Murolavoro. That's v-i-a new word m-u-r-o-l-a-v-o-r-o, number 17, that's one-seven. New line, 56127 Pisa. New line, Italy.

## 3.6 When spelling out telephone numbers, read each digit individually

Read each digit individually. Thus to say 113 4345, you would begin by saying *one one three* rather than *one hundred thirteen*. When reading out landline numbers, it is generally best to separate the country code from the rest of the number. Below is an example of how to spell out 0044 161 980 416 71.

> zero zero four four—that's the code for England—one six one; nine eight zero; four one six; seven one

Note that some people say *oh* rather than *zero*.

Whether you are dictating or noting down an address or phone number, make sure you repeat them at least twice. Even people who speak the same language often make mistakes with numbers.

## 3.7 Consider sending a fax, rather than an email, confirming what has been said

If you think that your message may not have been understood, then ask for their fax number, and then fax the information to them. A fax is better than an email in this case, as understanding an email address is often considerably more difficult than understanding a simple number.

## Leaving a telephone message: Summary

- Learn the structure and typical phrases of a call where someone leaves a message.
- Speak clearly and slowly.
- Get the caller to confirm what you have said, to read back the information you have given them, to repeat any important spellings of names.
- Use the international alphabet to spell out difficult names.
- Practise reading out your email address in a way that is easy for your interlocutor to write down.
- When giving a number, say each number as an individual digit (e.g. *one-two-three* rather than *one hundred and twenty three*).

# 4 VOICEMAIL AND ANSWERING MACHINES

## 4.1 Use an appropriate voicemail

When someone telephones you, they generally expect you to be there and to answer their call. Instead, if they get put through to your voicemail their reaction will range from frustration to possible annoyance.

The caller will be certainly not be impressed when hearing a voicemail such as:

*Sorry, but I am away from my desk, please leave a message after the tone.*

From the above voicemail the caller cannot understand:

- whether you are at work today
- if you are at work, when you will be back at your desk

A caller will also be frustrated to hear a voicemail such as:

*I am not at my desk at the moment, I will be back in half an hour. Please leave a message after the tone.*

Half an hour starting from when?

If your company does not already have a policy for voicemail, then consider the following:

1. state your name and today's date
2. say when you will be next at your desk / available (say the time using GMT, CET etc)
3. provide an alternative number for the caller to reach you (e.g. your cell number, a colleague's internal number)

Thus a typical message would be:

*You have reached the desk of Adrian Wallwork on Wednesday, March 10. You can call me back after three o'clock central European time. For urgent calls, my cell number is: 0039 340 7888 3041. Or you can leave a message after the tone.*

Try to keep the message as short as possible. The above voicemail could be re-recorded as:

*Adrian Wallwork. I am free after three o'clock CET. My cell number is: 0039 340 7888 3041. Or leave a message after the tone.*

## 4.1 Use an appropriate voicemail (cont.)

It is particularly important to keep the message short if your voicemail is in English and in your own language. It is probably best to leave the first part of the voicemail in English: if your foreign clients hear your voice message in your own language they will probably put the phone down immediately. On the other hand, clients of your own language are more likely to presume that there will also be a message in their language. Clearly, however, this depends on the number of English-speaking clients you have and how important they are.

If you are leaving a voicemail to say that you are away for several days, then you can say:

> *You have reached the desk of Adrian Wallwork. I will be back at work on Monday, March 15. Either leave a message after the tone or contact my colleague Anna Southern on extension 24.*

By providing a specific voicemail, you give the caller clear options.

## 4.2 Leaving a message on someone's answering machine

When leaving a message on someone else's answering machine you give much the same information as when you leave a message with a switchboard operator (see Chapter 3):

1. state your name (and who you work for and in what capacity)
2. today's date and time (only if the voicemail has informed you that the person is away for more than a day)
3. the reason for your call
4. the number you want to be called back on (or email address)
5. the best time(s) to reach you

To learn how to leave your email address see 3.5.

Below are some examples:

> Hi. This is Peter Hall from Metafora, in Pisa, Italy. I'm ringing with regard to our invoice number 230 dated 14th December of last year, which is still outstanding. Could you email me? It's p for peter dot Hall, h-a-double-l at metafora, that's m-e-t-a-f-o-r-a, dot i-t for Italy. Thanks a lot bye.

> Adrian Wallwork, CEO at E4AC. Your number was given me by Vesna Gugurevic who told me you might be interested in collaborating on an EU project regarding machine translation. You can call me on 0044 340 7888 3041. In any case Vesna also gave me your email, so I will contact you via email too. Thanks. Goodbye.

> Hi, this is Kate. It's 10.00 a.m. on Wednesday the fifth. Could you call me on my mobile as soon as you get back. Thanks.

## Voicemail and answering machines: Summary

- Voicemail: your name and today's date; when you will be next at your desk / available (say the time using GMT, CET etc); an alternative number for the caller to reach you (e.g. your cell number, a colleague's internal number)

- Answering machine: name, company, position; date and time; reason for call; your number and best time to contact you

# 5 RECEIVING CALLS: WORKING ON RECEPTION / SWITCHBOARD

## 5.1 Initial salutations

Most company receptionists follow this procedure when receiving an external phone call:

1. state name of company
2. say 'good morning / afternoon / evening'
3. ask how you can help

Points one and two can be reversed. So a typical initial statement by the receptionist is:

> ABC Solutions. Good morning, how can I help you?

The idea is to keep what you say to the minimum but at the same time:

- provide the necessary information (the name of the company), then the caller knows that they have dialed the correct number
- be polite by saying 'good morning' and asking how to help

Some companies also ask their receptionists to give their names:

> ABC Solutions. Katrine speaking, how can I help you?

If you don't know the caller or have no well established relation with the caller, then the next step is to transfer the call (see 5.2). However, if the person is a frequent caller, you might like to say:

> Oh, good morning Chandra, how are you?

> Good afternoon Mr Mendez, what's the weather like in Lima today?

Having a friendly manner will help the success of any future interactions and will give the impression of a well-mannered approachable company. However, ensure that such friendly exchanges are appropriate and also that they do not waste the caller's time.

## 5.2 Transferring the call for a client

Most of a receptionist's job is to transfer the caller to the right person. In order to do this, you need to find out:

1. the name of the caller
2. the name of the person the caller wants to speak to
3. why the caller wants to speak to the person in your company

To do the first and the second, there are many phrases you can use:

*Can I have your name please?*

*Could you give me your name please?*

*Sorry, what was your name again please?* [If you didn't understand their name the first time]

*Who shall I say is calling?*

*Sorry, who did [you say] you want to speak to?*

The third point, i.e. finding out why the caller is calling, is important as it helps you to establish the right person for the caller to speak to. The caller may have the name of one person in your company, but in reality another person might be more suitable. You can say:

*Can I ask what it is about?*

*Could you tell me what it is in connection with?*

Then, when you hear the reason you can either transfer the caller to:

- the person requested
- another more suitable person in terms of competency
- a secretary or colleague of the person requested if you think the person requested is likely to be busy or will not want to be disturbed

If the person requested is available, typically you will say:

*I'll try to put you through.*

*I will try to connect you.*

*I'll just put you on hold while I try to put you through.*

If you want to transfer the caller to another person:

*Actually, the personnel manager is now Antoine Delon, I'll just put your through.*

*I'm sorry but Khalil is not available, I'll just put you through to his secretary.*

## 5.2 Transferring the call for a client (cont.)

If the person requested is not available, then you can say what the person is doing and when they will be available:

*I'm afraid that Jeanette is in a meeting at the moment. It should end at 11.30.*

*She's not in the office today, but she should be in tomorrow.*

Then you can give the caller a choice of next moves:

*Could you call back at 11.30?*

*Shall I ask him to call you back?*

*Can you send her an email?*

*Would you like to leave a message?*

From your point of view, the last option is the most difficult as it involves you understanding the message. So it is generally best to try the other options first.

## 5.3 Transferring the call for a colleague: informal version

If you work for a company that has offices in many countries, then you will probably receive phone calls from colleagues who will speak to you in English. In the dialog below the receptionist's role is in *italics*. The conversation is informal.

*ABC good morning, can I help you?*

Hi it's George from the Madrid office, can I speak to Marion please?

*Hi George, how are you doing?. Hang on a second. I'll just check if Marion is in.*

*[Talking to Marion] Marion, it's George from Madrid, do you want to speak to him or shall I tell him you are not in?*

*[Talking to George again] Hi there, sorry she is not at her desk at the moment, could you call back later please.*

Yes of course. Thanks. Bye.

The words and phrases that indicate informality are:

- *hi* (more formal = *good morning* or *hello*)
- *it's* - to indicate who you are (more formal = *this is*)
- *I'll* - contracted forms (e.g. *I'll* instead of *I will*) may be a sign of informality, although they are used in more formal English too.
- *Marion* - use of first names rather than last names
- *how are you doing?* (in a more formal context asking someone how they are is probably inappropriate)
- *hang on* - imperative forms without using *please*; in addition the verb *to hang on* is very colloquial (more formal: *please hold the line*)
- *hang on a sec* (i.e. *second*) - use of short forms of words
- *bye* is another short form (formal: *goodbye*)

## 5.4 Transferring the call for a colleague: more formal version

Below is a similar conversation to the one in the previous subsection, but this time more formal. The caller is in normal script, and the receptionist in italics.

> This is George from the Madrid office, may I speak to Marion Cosic please? // Could you put me through to Marion please?
>
> *Could you hold the line please, I will just put you through to Mrs Cosic. ... Hello, sorry to keep you waiting but Mrs Cosic is not at her desk at the moment. Would you mind calling back later please? // Would you like me to get her to ring you back?*
>
> If she could call me back before midday that would be perfect. Thank you for your help.
>
> *You're welcome and I will make sure that she gets your message.*

The dialog above also highlights that i) formal English tends to use longer expressions, ii) the receptionist is more explicit in saying what steps he / she will take (e.g. *I will make sure that she gets your message*).


## 5.5 Creating a friendly relationship with colleagues

The role of a receptionist or switchboard operator does not have to be limited to transferring calls or dealing with difficult callers (Chapter 8). The role can also be to create a friendly atmosphere that enhances positive communication. Here is an example:

> ABC Good morning.
>
> *Hi Gabry, it's Marina, how are you?*
>
> Fine thanks and you?
>
> *Everything is fine thanks! So what's the weather like in Madrid?*
>
> It has been raining all week.
>
> *Oh dear. That's too bad. Here in Warsaw it has been sunny since Monday. Listen, the reason I'm calling is ...*

The conversation above focuses on the weather, but other topics that are not too personal could include the weekend, holidays (past and future), sports events etc.

## 5.6 Choosing the easiest phrase to say

Whenever you make a phone call, use the phrase that comes the easiest and quickest to you. This means choosing a phrase that does not contain words that are difficult to say (due to pronunciation or stress) or difficult to remember. Normally the phrase that you will remember is likely to be either the shortest or the one that is most similar to the equivalent phrase in your own language.

For example, there are two typical phrases that can be used when transferring a call:

> I'll try to put you through.
>
> I'll try to connect you.

The first phrase is more commonly used by native speakers. However it is a phrasal verb: *to put someone through*. This means that it is:

- difficult to remember, the combination of verb and preposition (*put + through*) does not give much indication of the meaning
- more difficult to use: do you *put someone through* or *put through someone*?
- more difficult perhaps for the caller to understand

The second phrase is probably easier to use because:

- its meaning is clear and logical: it is a verb used in many other situations, whereas *put through* tends to be exclusively used in the meaning of transferring a call
- it may be similar to a word in your own language (particularly Latinate languages)

Your objective is not to demonstrate your incredible knowledge of English, but simply to communicate as clearly and as rapidly as possible.

At the end of this book is a chapter on Useful Phrases (Chapter 18). It contains several alternatives for saying the same thing. I suggest that in each case you choose the phrase that is the easiest for you to use and remember. However, you need to be familiar with the other phrases in case someone uses them with you - you thus need to be able to understand such other phrases as well.

## 5.7 Use of *will* and present perfect

The most frequently used tense by receptionists is the future - *will* ( *'ll* = the contracted form). Use *will* when you react to what the caller says to you or to say what you are going to do next.

*I'll just check if she's in her office.*

*I'll just see if I can find someone to help you.*

*I'll just put you through.*

*I'll make sure she gets your message.*

*I'll ask her to call you.*

*I'll just read that back to you. [After you have taken down a message]*

The word *just* is frequently used when you say what you are about to do now. It doesn't add anything to the meaning of the phrase, but sounds polite. However, you can omit it.

The other frequently used tense is the present perfect, to report what you have just done in response to the caller's request.

*I've just spoken to the marketing manager and she says ...*

*I've given her your message and she says she'll call you back in 10 minutes.*

## 5.8 Being proactive and helpful

How you handle a phone conversation at work can also improve business relationships between clients and customers. Be proactive - i.e. think in advance what the caller might need and thus to provide them with such information even before they ask for it.

Look at the conversation below between a caller and someone on reception. The caller's side of the dialog is in italics. Note how there are two versions (a, b) of the receptionist's side of the dialog (in normal script): the first is unhelpful, the second offers much more information and in a friendly tone.

*Could I speak to Desdemona Alvarez please?*

a. She's not at her desk.

b. I'm afraid she's not at her desk at the moment, I think she's just gone out to lunch.

*Do you know what time she will back?*

a. No.

b. Well she normally takes about half an hour. Shall I get her to call you as soon as she comes back?

*Could you possibly give me her mobile number. It's quite urgent.*

a. I am not authorized to.

b. I am really sorry but I am not authorized to.

*Well could you ring her for me and ask her to call me back?*

a. OK.

b. Certainly, I can do that for you. Could I have your name please?

*Could you ask her to call me, it's Penny Dalgarno.*

a. What?

b. Sorry what was your surname again? Could you spell it for me.

*Dalgarno. d-a-l-g-a-r-n-o.*

a. OK. Goodbye.

b. OK Penny I'll ring her straight away. Goodbye.

Even if you make more mistakes in your English, try to use Version B type responses. Version A responses will simply make the caller frustrated.

## 5.9 Adopting a friendly tone

Research has shown that if you try to smile while talking on the telephone and also imagine that the caller is also smiling, then your tone of voice will sound friendly and helpful.

If you sound friendly then the caller is likely to make more effort to help you to understand what he / she is saying. On the other hand, an indifferent or monotone voice may have a negative effect on the success of your communication.

To avoid a monotone you need to learn what words you can stress to make what you say more dynamic and less flat (see 9.6).

## 5.10 Taking a message

Here is an example of a conversation between a receptionist (in italics) and a caller. The problem is that the receptionist cannot understand the caller due to a bad line. She is thus having great difficulty understanding the message that the caller wants to leave.

It is important for you to understand that the same conversation could take place between two native speakers, and that the strategies that the receptionist uses are typical of those of a native speaker.

*I am sorry could you speak up a bit, I can't hear you very well.*

Sorry, is that better? I was saying that I would like to speak to the sales manager.

*To the what manager? Sorry.*

The sales manager.

*OK, sorry she's not in at the moment would you like me to take a message?*

Yes, could you ask her to ring me. It's Richard Gabbertas.

*Sorry I didn't catch your last name.*

Gabbertas, that's g a b b e r t a s. And I'm calling from meta4.

*Could you spell that for me please?*

Yes it's m e t a and then the figure four, have you got that?

*The figure for what sorry?*

Sorry what I mean is that it is meta in letters and then the number 4. And my number is 020 8347 1254, and I am calling with regard to distribution problems in the London area.

*Listen, I am extremely sorry but I think it might be best if you sent her an email. Do you have her address?*

Note how the receptionist realises that there is no point in continuing the conversation as the caller is likely to become frustrated. The caller's frustration could lead to a negative impression of the receptionist's company. If you are having difficulty understanding, then often the simplest solution is to ask the caller to send an email. This means that you have to:

- be able to give an email address over the phone (see 3.5).
- learn how to take a message (see the dialog in 3.1).
- learn what to do if you don't understand (see Chapter 14).

## 5.11 Dealing with wrong numbers

Here are two examples of how to deal with people who have dialed the wrong number. The person who answers the call is in *italics*. In the first example, the caller has dialed the wrong internal number.

*Claire Henson.*

Could I have invoice processing please?

*I'm afraid you've come through to production.*

Sorry I must have dialed the wrong number.

*Not to worry, I'll try and transfer you. Just hold on a sec.*

In the next example, the caller has dialed the right number, but the person requested no longer works there.

*ABC. Good morning. Can I help you?*

Could I speak to Augustin Miquel please?

*To who? Sorry.*

Augustin Miquel.

*Sorry, but I think you must have dialed the wrong number. No one by that name works at ABC.*

Can you just double check? I was given this number: 0044 208 546 9876

*Well, that's the correct number. Just hold the line, and I'll check for you.*

*Sorry, you do have the right number but Augustin Miquel no longer works here. And I am afraid we don't know where he currently works.*

OK, thanks for your help and sorry to have bothered you.

## Receiving a call: Summary

- If you are on the switchboard and you receive a call: state the name of your company, say 'good morning', and ask how you can help.
- Before transferring a call, find out: name of caller, name of person caller wants to speak to, why caller is ringing.
- If person requested is not available, tell caller when the person will be available and / or how to contact the person via email or mobile.
- Choose the right level of formality.
- Choose the easiest phrases to say.
- Use *will* to say what you are going to do next, and the present perfect to say what you have done.
- Be proactive, helpful and friendly.
- If you are having difficulty understanding, ask the caller to send an email.
- Ask the caller to call back if you feel that you are unable to respond effectively.

# 6 FINDING OUT ABOUT ANOTHER COMPANY, GIVING INFORMATION ABOUT YOUR COMPANY

## 6.1 Responding to a caller who wants information about your company

If you work in sales or marketing, you may be contacted by potential clients who are interested in knowing something about your company. Below is a conversation between a caller (Maria, a potential client) and a marketing manager (Pedro, in italics) of a company called MF Consulting that produces software for financial markets. The square brackets contain an explanation of what Pedro is doing at each stage of the exchange.

This is Maria Bragança from the Banco di Finanças in San Paolo, Brazil.

*Good morning Ms Bragança. Thank you for calling. How can I help you?* [Repeats name of the client, this gives him the chance to check he has heard correctly. He immediately tries to create a professional impression by being polite]

Your company was recommended to us by Augustin Evangelista, who I think you know.

*Oh Augustin yes, we are currently working on a project for his company.* [Uses this opportunity to show that his company is working on an important project for someone who is known to the potential client]

Well I'd just like a few details about how your company operates.

*First of all, let me introduce myself, I am Pedro Gonzales and I am the marketing manager at MF Consulting.* [It is important for the client to know who she is talking to, and that this person is the most appropriate person to be talking to]

Pedro, pleased to meet you.

*Well, as you probably know MF Consulting works in the field of the electronic money markets here in Argentina. We not only develop products but also projects, so we work specifically for companies in creating dedicated products for them. Our clients are both banks and financial institutes such as...* [Gives details of what MF do and who their clients are. Mentioning these clients gives more credibility to his company]

*We work both in and outside South America and we have offices in London too. In fact I am calling from the London office now. Do you have operations in Europe?*

## 6.1 Responding to a caller who wants information about your company (cont.)

*MF Consulting also has offices in London, but not under the name of MF Consulting but under the name of MFQ which is a company partially owned by MF Consulting. So perhaps it might be more convenient for you to meet at our offices in London, which are located in...* [Tries to think from the point of view of the client, what will be the easiest for her?]

OK, that's actually very close to our offices.

*Really? That's perfect. Well, you could contact Amélia Silva who works at MFQ, London, I think she also speaks Spanish.* [Shows a positive reaction to what Maria says. Then gives Maria another good reason for contacting Silva – she speaks the same language as Maria]

Well that would be useful.

*Well if you could send me an email specifying exactly what it is you want to know. Then I will forward your email to Amélia and put you in cc.* [Asks for further details, so that he can meet Maria's request for information more precisely. Shows that he will keep Maria informed at each stage of any future communication]

OK, I'll do that right away.

*Shall I give you my email address? It's... When I get your mail, I'll send you more information about the company and our field of activity. I'll also send you the addresses of our two websites. On the MF Consulting website you can find first level information about our products, and on the MFQ website you'll find details of a particular product called MFQuote which I think you might find interesting.* [Gives more pertinent information that he thinks Maria might be interested in]

OK, that sounds good.

*OK brilliant. Do you have any more questions that you would like to ask now?*

No I think that's everything for the moment. Thanks for your help.

*Well, could I just ask you a few questions before we end the call?* [To be able to satisfy the client's requests, Pedro needs to know more details about Maria's company]

Sure go ahead.

Clearly Maria, the person asking the questions, is likely to talk less than Pedro, the person answering the questions. However Pedro tries to be reasonably concise in his answers and thus does not dominate the conversation too much. This is important, as both parties in the conversation need to feel equally important. Pedro reacts positively to Maria's comments (e.g. the fact that Maria's office in London is close to MFQ's office). Also, he tries to balance the talking time spent by each person by asking Maria about her company.

## 6.2 Calling a company to find out information about that company

If you are calling another company to find out details about them, you might also like to consider the following:

- Check that you are talking to the most appropriate person.
- Find out if you have any contacts in common (you can then use these contacts to make any verifications on the company).
- Ensure you get the person's contact details.
- Ensure that the person knows what you expect him / her to do next.

## 6.3 Calling someone in a company to make a cold sale

'Cold calling' is the term used for a salesperson who makes an unsolicited (i.e. not specifically requested) call to a company (for reasons of simplicity I will refer to the person contacted as a 'prospect') in the hope of getting the target company interested in a product or service.

This is possibly the most difficult kind of telephone call to make, the key element is to make a positive impression and to leave the prospect wanting to know more about your product or service.

A possible 8-step strategy is:

1. Give your name and the name of your company. Check that you are speaking to the right person.
2. Ask if this is a good moment to speak, and state how long the call will take.
3. Explain briefly what your company does.
4. Underline the potential benefit for the prospect of working with you and your company.
5. State what you plan to do next.
6. Find out more about the prospect's company and stress again how your company can meet their needs.
7. Repeat your name and the name of your company.
8. Thank the prospect for his / her time.

Below is a dialog that illustrates this strategy. The whole call should not last more than three or four minutes. You want to leave the prospect thinking that his / her time has not been wasted. The numbers in square brackets after each piece of dialog correspond to the eight steps of the strategy outlined above.

## 6.3 Calling someone in a company to make a cold sale (cont.)

*This is Takeshi Sato and I am calling from Takagi Consulting in Tokyo. Am I talking to John Smith in marketing?* [1]

Yes.

*Is this a good moment for you to speak? I won't take up more than three or four minutes of your time.* [2]

OK. But keep it as brief as possible, I have a meeting in a few minutes.

*Takagi Consulting helps companies like yours sell their products in the Japanese and Korean markets. We provide not only contacts but also short training courses in Japanese and Korean culture for your staff.* [3]

I see.

*Given that I have heard from Damo Suzuki [third party known both to Smith and Sato] that you are planning to launch your products in... I thought you might be interested in... Many European and US customers find the Japanese and Korean markets difficult to break in to, and we believe that we can fast forward negotiations by up to two months, thus saving you considerable costs.* [4]

OK.

*So the reason I am calling is to find out whether it would be OK to send you some of our documentation, and then to call you again some time next week.* [5]

That sounds fine.

*Just so that I can confirm what Damo told me, could I just ask you two more questions.* [6]

Go ahead.

*First... Second.*

Blah blah.

*OK, that's brilliant, you have been very helpful. So as I said, my name is Takeshi Sato and the company is Takagi Consulting. You can expect an email from me in the next hour. Can I just check your address?* [7]

Sure.

*So it's j.smith@smith.com*

That's right.

*Well, thank you so much for your time. And I am sure you will find our offer interesting.* [8]

The dialog above illustrates the most optimistic outcome. Frequently, you will find that your prospect is not interested, in which case you can say that you will call back in a few months to see if the position has changed. You can also mention that in any case you will send him / her an email giving more details.

## Finding out about another company, giving info about your own company: Summary

When making a sales call ensure you create a professional impression by being polite, introduce yourself clearly, use opportunities to highlight the benefits of their company to work with yours, mention names of your clients to give your company credibility, and be positive in your reactions to what the other person says. Follow these steps:

- Give your name and the name of your company. Check that you are speaking to the right person.
- Ask if this is a good moment to speak and state how long the call will take.
- Explain briefly what your company does.
- Underline the potential benefit for the prospect of working with you and your company.
- State what you plan to do next.
- Find out more about the prospect's company and stress again how your company can meet their needs.
- Repeat your name and the name of your company.
- Thank the prospect for his / her time.

# 7 CHASING

## 7.1 Chasing a payment

'Chasing' means contacting someone to find out at what point they are with a certain task, which you think should have already been fulfilled. In other words, this other person is behind schedule and this fact is making life difficult for you or your company.

One of the most common forms of chasing is with creditors who have failed to pay your invoice. Below is a typical conversation, the caller (i.e. the person who wants payment) is in *italics*.

Good morning, ABC.

*Is that invoice processing?*

It is yes.

*This is Josef Polache. I'm ringing from a company called Dragos in Moldova. We have three outstanding invoices with your company.*

Just one moment and I'll call up the account. 'Dragos' did you say? D-R-A-G-O-S?

*Yes that's right.*

I won't keep you one moment. Sorry to keep you waiting. The only one the system is showing as outstanding is Invoice 230, dated 16 December. We have a record of that invoice on our system but as yet it's not been presented for payment. Can you give me the other two invoices please?

*The other two are numbers 40, which was sent to you on January 13, and 190, which was sent on March 30.*

OK just one moment. I'm afraid we don't have a record of either of them on our system, would they have been sent to someone's particular attention that you know of?

*Right, number 40 was sent to 'Business Technology', and 190 to Hilary Cummings.*

Could you possibly fax me a copy of those two invoices?

*Yes of course.*

My name's Lee.

## 7.1 Chasing a payment (cont.)

*Lee? L double E*

Yes, And the fax number is 01661

*yep*

721

*yep*

2077

*2077. Can I have your surname as well please?*

Yuldiz. Y-U-L-D-I-Z. And I'll put you through to that department because they are actually holding on to the first invoice which as I said has not yet been presented for payment, but they will be able to explain better as to why it's still outstanding.

*Which one is that? 230?*

230. That's right yeah.

*So you're going to put me through to who sorry?*

A gentleman called Jamie Gallagher. I'm sorry I couldn't have been more help myself.

*Not to worry.*

Your call has been forwarded to an automatic voice message system. "Jamie Gallagher" is on the phone. At the tone please record your message. When you have finished recording you may hang up. Or press 1 for more options.

The dialog above highlights that you might need to give a lot of details and even then not get the information you required. In fact the caller gets put onto an automatic voice message system. Let's imagine that the caller now tries ringing back.

Good morning, ABC.

*Could you put me through to Jamie Gallagher please.*

Who shall I say is calling?

*This is Josef Polache. I'm ringing from Dragos in Moldova.*

Sorry, what was your name again?

*Josef Polache*

From Dragon?

## 7.1 Chasing a payment (cont.)

*No, Dragos–G-O-S*

OK, I'll just put you through.

Jamie Gallagher.

*I am calling from Dragos in Moldova with regard to two invoices that are still outstanding.*

Just bear with me two secs. Dragos?

*Yes, Dragos.*

From Moldova? Is that Moldovan leus?

*Yes, but the invoices are actually in euros not leus.*

Have you got the amounts?

*Yes, the first, invoice number 10, is for 10,500 euros and the other, invoice number 27 is for 987. euros.*

That's actually been paid. It was paid on the eleventh of February.

*Which one?*

Invoice 27.

*What about invoice number ten?*

I can't see anything on here. # What I do if I receive an invoice is to log it on the system, so that we've got it. And then I send them out to the contact to get authorised.

*OK.*

Sometimes there can be a bit of a delay.

*\* A delay, OK. In any case, I'll email invoice number ten again.*

# Yes, then what I'll do is I'll actually phone the contact. Do you have a name internally at ABC?

*\* You mean the name of the person who we deal with?*

Yes.

*Yes, I'll put the contact name in the email.*

And if you want to write your number on there as well then I'll be able to contact you and talk to you later on today.

*\* OK, I'll put my telephone number. Thanks a lot then goodbye.*

Cheers.

## 7.1 Chasing a payment (cont.)

The two dialogs above are based on transcripts of real telephone calls where the person in invoice processing is a native English speaker. What they highlight is that:

- the native speaker may speak a lot and give you a lot of detail that may not even be useful for you (these phrases are marked with a #). In such cases, the speaker will speak faster than usual to indicate that what he / she is saying is of less importance
- it helps if you continuously check the information you are being given – see the phrases in the dialog that begin with an asterisk (*)

## 7.2 Chasing an order

When you place an order with a company and there is a delay in receiving the order, you will normally need to provide the following information:

- the order number (or what the order was for)
- when the order was placed
- the name of the person or department the order was placed with
- when the order was originally expected
- how many times, if any, you have already chased this order
- your final deadline for receiving the order

It might also help to prepare something to motivate the receiver of your call to act urgently by thinking of possible consequences if they fail to deliver the order by the requested time. Here is a typical dialog (the caller is in *italics*):

*I am calling about our order number 213, dated 17 September, for...*

Yes, what can I do for you?

*We were expecting delivery on the first of October. We then called you on the third of October, and you assured us that we could expect delivery by the end of that week. It is now October 17, and we still haven't received the order.*

I am very sorry to hear that. Can you give me the name of the person who you received confirmation of the order from.

*Yes, it was Tyler Watts. But listen, if we don't receive the order in the next two days, I am afraid we will have to cancel it.*

I will call Mrs Watts now and ask her to call you. Does she have your number?

*Yes, but I'l give it to you anyway it's.*

## 7.3 Chasing a document, report etc

When you are chasing a document that someone was supposed to have written (i.e. you asked them to write it and so far they have failed to do so), typically the other person is someone within your company. This does not automatically mean that you can be angry with them. If you want to motivate someone to do something for you, be firm but polite. Obviously, your manner will also depend on your position in the company hierarchy with respect to the other person's.

Here is a possible dialog. The person who made the request, who is higher up in the hierarchy, is in *italics*.

*I was wondering how you were getting on with the sales report.*

Well, I am afraid I still haven't had time to finish it yet.

*OK, I understand that you are under a lot of pressure at the moment, but the report was due in at the beginning of last week. Realistically, when do you think you can finish it?*

I would hope by the end of this week.

*Well, could we make that a definite deadline? I would like to receive the report not later than 16.00 on Friday this week.*

OK.

*I am counting on you to do this, because I promised the sales manager that I would give it to him on Monday morning, and I need a couple of hours to go through it first. So it is very important that you get it to me on time.*

OK, I'll do my best. And I am sorry you have had to wait so long for it.

*That's OK. If you could just do everything you can to get it to me on Friday. Thank you.*

Although the person making the request is higher up in the hierarchy, note how she:

- is never rude or threatening
- makes her initial question very soft by beginning with *I was wondering*, which is typically also used when requesting a favor
- makes her request sound like a joint responsibility by using *we*, rather than *I* or *you* (*could we make that a definite deadline*)
- sets a specific deadline and elicits agreement on this deadline
- stresses the importance of the request and the consequences for her if the request is not met – this means that the other person fully understands why he / she needs to meet the request

## 7.3 Chasing a document, report etc (cont.)

Here is the same situation, but this time with two colleagues who are on the same level in the company hierarchy.

*I was wondering how you were getting on with the sales report.*

Well, I am afraid I still haven't had time to finish it yet.

*Do you have any idea of when you might be able to finish it?*

I would hope by the end of this week.

*OK, if you could possibly get it to me by 16.00 on Friday this week, that would be great. The thing is I promised the sales manager that I would give it to him on Monday morning, and I need a couple of hours to go through it first. It won't look good on me if I am late with it, as I am sure you will understand.*

In this case, the person making the request does not have the authority to put any great pressure on the report writer. In this case, she can only make the report writer try and understand the difficulty she will be in if she doesn't get the report in on time. Getting the other person to empathize with you is generally a good tactic.

For more on adopting a soft approach see 9.1.

## Chasing: Summary

- Gather all the info that you might need before you make the call e.g. invoice and order numbers and dates, who invoices and orders were sent to, what progress (if any) you have made so far in getting the info you require, when your deadline is.
- Be prepared for a lot of questions from your interlocutor and also for them giving your a lot of irrelevant information.
- Try to motivate the person you are chasing so that they will carry out your request – be firm and polite, and tell them what the consequences are for you and for them if they don't help you quickly (make requests sound like a joint responsibility).
- Never be rude or threatening.

# 8 DEALING WITH DIFFICULT CALLERS AND UNHELPFUL STAFF

## 8.1 Dealing with people who are trying to sell you a product / service that your company is not interested in

If you work on reception, you will inevitably get a lot of calls from companies or individuals trying to sell your company something that you know from past experience is of no interest to the company.

You can waste a lot of time on such callers with the added disadvantage that while you are dealing with them other callers will have to wait.

Here are some phrases you can use to discourage such callers.

*I am sorry but we already have a provider of such a service / product.*

*Sorry, but we are not interested in such a service / product.*

However, the caller may still insist on being put through to someone else in the company. For such people you can say:

*I am sorry but I am not authorized to transfer your call. Goodbye.*

*I suggest you go on our website, go under 'contacts' and send us an email. Thank you for calling. Goodbye.*

If this does not work, then you need to close the call. But don't simply put the phone down, announce to the caller that you are going to finish the call.

*Listen, I am really sorry, but I have to terminate our call now. Goodbye.*

## 8.2 Dealing with people who are waiting for a response from someone within the company but have had no reply

Companies receive a lot of written correspondence from people who are trying to sell something or who are seeking a job. Much of this written correspondence may go unanswered. Consequently, the people who have sent the letters and emails decide to telephone to get an update on their situation. If you are on reception, your job may involve trying to stop these people from talking directly with the person to whom they addressed their letter or email. Here is a typical conversation. The receptionist is in *italics*.

*ABC. Good morning. How can I help you?*

I am calling to the know the status of my CV.

*Did you send the CV in response to a specific advertisement?*

No.

*When did you send the CV?*

Last week.

*Well I suggest you wait another two to three weeks, and if you have heard nothing you can assume that the company has no suitable positions available.*

But could you put me through to the human resources department please?

*No, I am sorry but I am not authorized to do so.*

But this is urgent.

*I am really sorry but I have another caller waiting. Thank you. Goodbye.*

The same kind of caller will try to get past the receptionist by asking for the human resources manager by name, as illustrated by the next extract:

*ABC. Good morning. How can I help you?*

Could you put me through to Kimmochi Higashikuni in HR please.

*Can I ask what it is about?*

I sent him my CV a few weeks ago and I have heard nothing.

*Well, I am sorry but I am not authorized to transfer your call. I can assure you that you will be contacted if there is a position available for you. Thank your for your call. Goodbye.*

The key here is the phrase *Can I ask what it is about?* Without asking this question you risk transferring a call that is going to waste your colleague's time. Also, by saying *goodbye* you indicate that the call is about to be terminated.

## 8.2 Dealing with people who are waiting for a response from someone within the company but have had no reply (cont.)

You can use the same tactics with other callers who try to get past the switchboard by asking to speak to someone directly. As an alternative you can also say:

> I suggest that you contact Mrs X by email.

And when the caller says that they have already tried to do this but with no success, you can say:

> Well, I know Mrs X is very busy at the moment. I am very sorry. Goodbye.

## 8.3 Switchboard: dealing with a client who wants to register a complaint

If you are on reception and you receive a call from a client who wants to register a complaint regarding the company, it is imperative that you deal with the client in a calm empathetic way. Here are some key phrases you can use:

> I am sorry to hear that [i.e. about the client's problem], I will just put you through to someone who can help you.

> To ensure that I put you through to the right person, can you just tell who you have been in contact with before?

## 8.4 Person responsible: dealing with a client who has received poor service

Below is a conversation between a man (in *italics*) who is responsible for delivering a service / product to a female client, however there has been a significant delay in delivery. Note how the provider of the service:

- empathizes with the client by showing that he understands the difficulties he is causing her
- shows that the client is valuable to his company
- explains the problem and reassures the client that he is doing everything he can to resolve the situation
- gives the client a deadline for the receipt of the service / product
- apologizes again and reassures again

The fact is that you promised delivery last week.

*You are absolutely right, and I can imagine that this is causing you a lot of problems.*

Yes, we were depending on the order.

*Well, the problem is that the order was inputted incorrectly into our system. This is entirely our fault and I apologize for this. But the good news is that a new order is being dispatched today.*

So when does that mean we will receive it? We need it by Thursday at the latest.

*Well the order has been given top priority and you should certainly receive it by Wednesday.*

OK.

*We very much value your custom so we are happy to give you an extra 10% discount in addition to the 10% you are already receiving. I know that this cannot compensate for the problems we have caused you, but I hope it shows our appreciation of your patience. [This kind of compensation is in reality quite rare!]*

OK that sounds good.

*We will also try to ensure sure that this never happens again. In any case, once I again I hope you will accept our apologies.*

## 8.5 Dealing with rude callers

When someone is impolite to you over the phone, the best solution is to adopt a friendly but firm tone and not to enter into any arguments. If someone starts shouting or insulting you, you can try to calm them down by saying one or more of the following:

*I understand that you are feeling very frustrated. Can I ask you just to be patient a little longer while I find someone who can help you.*

*I appreciate that the delay in receiving the goods is causing you considerable problems. I am doing my best to solve the problem for you.*

*I apologize for the inconvenience this is causing you. I assure you that our technical staff are doing all they can to resolve the problem. Would you like to speak to one of the technicians?*

## 8.6 Dealing with unhelpful staff when you are the caller

If you are the caller, one of the most frustrating aspects is when you have to deal with someone who only gives you yes / no answers and seems reluctant to give you the information you are looking for. Here is a typical example:

*Could I speak to Desdemona Alvarez please?*

*She's not at her desk.*

*Do you know what time she will be back?*

*No.*

*Could you possibly give me her mobile number. It's quite urgent.*

*I am not authorized to.*

*Well could you ring her for me and ask her to call me back?*

*OK.*

In the above case, given the unhelpful nature of the person who answers the call, you cannot be sure that your message will be passed on. The solution is probably to ask to talk to a colleague or secretary of the person you wish to contact. So you can say:

*Could I speak to Desdemona Alvarez please?*

*She's not at her desk.*

*Could you put me through to someone in her department please.*

*[The line is transferred]*

*Yasmin Sabina.*

*Good morning, I am trying to get hold of Desdemona Alvarez.*

*She's not in yet. She should be in at 10.00, would you like to leave a message?*

In this case, the caller is lucky as Yasmin is more helpful. But if Yasmin too is unhelpful, then another solution is to try to call a different number within the company. This time, you need to be more insistent:

*I wonder if you can help me. I have been trying for the last twenty minutes to get in contact with Desdemona Alvarez. The matter is very urgent. Please can you find someone who can help me.*

## Dealing with difficult callers and unhelpful staff: Summary

- If you are a receptionist, devise ways of politely terminating a telephone call with unwanted callers e.g. by telling them you already have a provider of the product or service you are offering, by saying that you are not authorised to transfer the call to the person they want to speak to, by telling them to write an email.
- Empathize with callers who are registering a complaint. Show understanding and try to keep the caller calm. The same applies with impolite callers – the important thing is never to be impolite yourself no matter how insulting the caller gets.
- If you are the caller, and you fail to be able to be transferred to the desired person, ask to be transferred to that person's secretary or another person in their department. Again, don't be rude as this will not improve your chances of being transferred to the right person.

# 9 IMPROVING YOUR TELEPHONE MANNER

## 9.1 Avoid being too direct

When on the phone, you lose a substantial part of your communicative impact because the other person cannot see the expression on your face, nor can they read your body language. So make an extra effort to be polite and diplomatic. If you use very direct language, you may be in danger of sounding rather rude.

The second column in the table below shows some 'softer' versions of the phrases in the first column.

| POSSIBLY TOO DIRECT | SOFTER |
|---|---|
| Why didn't you send us the report earlier? | I was wondering why I hadn't received the report earlier. |
| You were supposed to be sending me a report. | Sorry but I was expecting to receive a report. |
| I need the revisions by tomorrow. | I was wondering if you could send me the revisions by tomorrow. |
| | Would it be possible by the end of the morning? |
| | Do you think you could send me the revisions by lunchtime today? |
| I'm calling you about ... | The reason I'm calling you is ... |

The 'softer' versions require more words and more complex grammar (e.g. use of conditionals, and the past continuous and past perfect). However, even if your English is low level, you could simply learn the phrases as if they were idioms rather than as grammatical constructions.

## 9.2 Help the person that you want to speak to

The person you wish to speak to may not be expecting your call. When they receive your call they will probably be in the middle of doing something else and need a few seconds to reorient themselves. To help them:

- say who you are – first and second name (the switchboard operator may not have given your correct name to your interlocutor)
- explain the context – i.e. the relevant communication you've had with this person, for example, *you may remember that I sent you a document two weeks ago, well I am calling because ...*
- explain why you are calling

If you are talking to someone with a low level of English, frequently summarize what you say. Ask questions to make sure your listener has understood. Don't merely say 'OK?', 'Have you got that?' because even if they say 'yes' it doesn't mean that they have necessarily understood.

If they don't understand what you are saying and you are speaking reasonably slowly, it may simply be that they don't understand a particular word or phrase. Try and re-phrase what you've said, instead of repeating exactly what you've said before.

## 9.3 Speak slowly and clearly

You may be nervous about making the call. When we are nervous we tend to speak fast, particularly at the beginning of a conversation. In any case it is best to speak slowly and clearly, with pauses between each piece of information. This is because:

- we give our interlocutor important info (our name and institute, and why we are calling) at the beginning of a call
- our interlocutor is probably still thinking about what he / she was doing before you interrupted them

Also, you can offer to repeat things:

> Would you like me to repeat my name again?

## 9.4 Don't be afraid to interrupt and make frequent summaries of what you think you have understood

If you are talking to a native English speaker, it may help you to remember that communication is two way. The responsibility for the fact that you do not understand something lies with your native-speaking interlocutor too. Your interlocutor will also want you to understand what he / she says. So keep checking that you have understood what is being said, and make summaries of your understanding so that your interlocutor can check whether you have correctly interpreted what he / she has said. See Chapter 14 to learn these skills.

## 9.5 Compensate for lack of body language

Unless you are using Skype (or equivalent) the other person will not be able to see the reaction on your face, or whether you are nodding in agreement or not. You can compensate for this by using one of the following expressions: *I see, yes, OK, right* or noises such as *a-huh* and *mmm*. Using such expressions also shows your interlocutor that you are still on the phone and are absorbing the information being given to you.

## 9.6 Learning to sound authoritative and competent

Given that your interlocutor cannot see you (1.5), the sound of your voice is imperative. You need to practise sounding confident and authoritative. You can do this by trying to imitate the voices of authoritative people on the telephone. There are many books written for classroom use for non-native speakers on telephoning skills. Search for books by EFL (English as a foreign language) publishers such as OUP, CUP, Macmillan and Longman.

These books contain CDs with scripts. You can practise imitating the tone and intonation of the speakers, by listening to the dialogs and reading the scripts, and then repeating what each person has said.

You can also find such scripts and dialogs in Business English books (published by the same publishers listed above), including my own books for Oxford University Press (OUP): *Business Options*, *Business Vision* and *International Express*.

Another option is to search for 'telephone skills' on YouTube.

Note how authoritative people speak. They tend to speak:

- quite loud (but not too loud)
- quite slowly
- very clearly
- in a way that shows that they believe in what they are saying (by sounding positive and enthusiastic)

Clearly, achieving a professional tone is not simple in a foreign language, but it can be done!

It also helps if you look professional too. If you sit upright, your voice will sound clearer.

Some experts believe that if you are dressed professionally (even though your interlocutor can't even see you), it will help you sound professional.

## 9.7 Evaluate your performance

You can improve your performance on the telephone by constantly analysing how effective your last phone call was. Think about the following:

1. Did I answer the phone / announce myself in an appropriate and polite way?
2. Did I handle the call in a professional way? Did I present a good image of my organization?
3. Was I friendly but competent at the same time?
4. Did I understand the interlocutor's questions? Did I interpret them correctly?
5. How did I manage any misunderstandings?
6. Did I speak slowly and clearly?
7. Will my interlocutor have had the impression that I was a cooperative and competent person?
8. Did I show any signs of frustration or annoyance?
9. Did I make frequent summaries to check that I had understood?
10. Did I make a final summary of the action points to be taken?

However, you are probably not the best person to assess your own performance.

If you can, ask a colleague to sit next to you while you are making a call in English and ask for feedback on the points above.

## Improving your telephone manner: Summary

- Avoid being too direct – use a soft approach.
- Always help your interlocutor to understand exactly what you want by expressing yourself slowly, clearly and unambiguously. Say who you are, the context (i.e. any helpful background info), and the reason for calling.
- Interrupt and make frequent summaries.
- Show that you are listening by making appropriate noises (*uh huh*) or by saying *yes, right* etc.
- Learn how to sound competent and authoritative by listening to scripts of telephone recordings.
- Get a colleague to evaluate your telephone performance.

# 10 WORKING ON A HELPDESK: KEY ISSUES

## 10.1 Do not panic. Listen to the full explanation before reacting

Companies often underestimate how important their helpdesks are. If a customer is regularly left unsatisfied by the helpdesk, then they are likely to change product or service providers.

Most customers use a helpdesk as a last resort. They have tried to manage the problem by themselves, they have tried the handbook / manual of the product, and their last hope is the helpdesk. And, they may need a solution urgently, so they tend to be already quite anxious when they call.

Unfortunately, native English speakers tend to find their experience with non-native helpdesk operators to be very frustrating. This chapter outlines some of the key issues and how to resolve them.

One basic cause of frustration for the customer is when the helpdesk operator fails to understand what the real problem is and focuses on something else.

For example, let's imagine a customer calls the helpdesk of an IT company because she has a problem with a software application. The customer (in *italics*) begins by saying:

> *I'm getting error message 30 and.*
>
> OK, error message 30. I understand, let me just look that up.

The problem is that the helpdesk operator has understood the words *error message 30* and automatically assumes that this is what the customer is calling about. Instead what the operator should have done is illustrated below:

## 10.1 Do not panic. Listen to the full explanation before reacting (cont.)

*I'm getting error message 30 and.*

OK, error message 30.

*But I have already looked that up and the explanation given did not solve the problem. So clearly there is another problem that is generating this error message.*

So let me just check that I have understood the problem. You get error message 30. You've checked the solution for message 30, but the problem remains. You think there must be another problem. Is that correct?

*That's correct.*

OK, well what we can do is.

Before you start trying to solve a problem, you need to check exactly what problem needs to be solved. Because you are speaking in a foreign language and because you have difficulties understanding native speakers, you may tend to panic when a native-speaking customer calls. This panic causes you to act on the first thing that you understand – in our case, *error message 30*. Instead you should:

1. let the customer explain exactly what the problem is – never interrupt to solve the problem before they have finished their explanation

2. at appropriate moments during the customer's explanation, repeat back what you think you have understood

3. at the end of the explanation, make a summary of what you have understood and wait for the customer to confirm that you have understood correctly

## 10.2 Admit that you have not understood

A similar problem to the one outlined in 10.1 is when operators do not admit that they have not understood. On the basis of a few key words that they think they have heard, they then draw the wrong conclusion.

So instead of just picking up key words, you need to ensure that you have got the complete sense. If you don't understand something, then it is imperative that you:

- admit that you don't understand rather than guessing
- use strategies to clarify meaning

To learn what to do and say when you don't understand the caller, see Chapter 14.


## 10.3 Improve your pronunciation

Another major problem is that the native speaker simply cannot understand what the operator is saying. This is generally due not just to poor English, but much more frequently to pronunciation problems.

It is critical to the success of your helpdesk line, that every operator learns to pronounce correctly the key words they need in order to give helpful explanations to customers.

To learn how to improve your pronunciation, see Chapter 16.


## 10.4 Ask the caller to speak more slowly

People call the helpdesk to get a problem solved quickly. They use the phone rather than email because they are hoping to get an instant response. Often they are under pressure from their bosses to solve the problem immediately. This means that although they know they should speak slowly and clearly, they are overtaken by the urgency of the situation and tend to speak fast. Another reason is that by speaking fast they hope to get their problem solved more quickly.

If the caller is speaking too fast and you cannot understand, then there is no benefit to either of you in pretending that you can you follow what they are saying. You need to say something to slow them down such as:

> *I appreciate that this is an urgent problem for you. But to solve it I need to understand exactly what the problem is. If you could speak more slowly this will really help me to identify your problem.*

## 10.5 Check whether the caller has a single problem or a multiple problem

Occasionally callers contact a helpdesk to resolve multiple problems. However, the helpdesk operator may not be clear that there are multiple and separate problems, and think that they are all part of one major problem.

Again the key is to constantly ask for clarification. You can say:

*Can I just check whether you have one specific problem or whether you are talking about several distinct problems?*

*I am really sorry, but I think I am getting a bit confused. Can we just try to deal with one problem at a time?*

## 10.6 Prepare possible customer questions and solutions to these questions

Callers expect to be given explicit replies to explicit questions. If you work on the helpdesk, brainstorm your team and think of all the possible questions that callers can ask you. I suggest that you keep a log of all these questions and write down clear solutions to these problems in English.

Every time there is a new problem, keep a record of what the problem was and how it was resolved. Then think of the best way of expressing the solution to this problem in English.

It is a good idea to get a native English speaker to check all the explanations that you give and that your pronunciation of any key words is correct.

## Key helpdesk issues: Summary

- If you work on a helpdesk, do not react to the first words that you understand. Instead wait for the caller to finish explaining their problem.
- Improve your pronunciation, particularly of key words that you are likely to hear and use during calls with customers.
- Admit when you don't understand something. Ask for clarification.
- Most customers will speak fast and maybe quite irate. Get them to slow down, by speaking calmly and slowly yourself. If necessary, ask the customer to speak more slowly.
- If the customer asks multiple questions, check whether these questions relate to the same or different problems. Then deal with each question separately.
- As a team, try to note down the solutions to typical problems as well as the English language constructions and vocabulary that you will need when you outline the solutions to the customer.

# 11 HELPDESK: DEALING WITH CUSTOMERS

## 11.1 Dealing with a customer's problem: a ten-step solution

Below is an example of a client who has a problem connecting his computer platform to a provider's website. In the dialog the part spoken by the helpdesk is in *italics*. As you read the dialog do not focus on the technical language, instead focus on the various questions that the helpdesk operator asks and the steps that he takes.

*ABC, good morning, Carlos speaking. [1] How can I be of assistance to you? [2]*

Hello I am calling from XYZ about a connection problem.

*OK can you give me your identification number and your name to update our system? [3]*

Sure, it's ................. The reason I am calling is that we have a problem with the connection of our platform to the B market.

*OK, let's see what we can do, I am sure we will be able to sort this out – what kind of problem is it exactly? [4]*

We noticed that there was a delay from the transmission data protocol at 8 o'clock this morning. Could you verify that this is correct because we didn't get any response from your platform.

*So, you didn't get any response from our platform. [5]*

*Could you try to connect to this IP address and port using the telnet command? [6]*

OK just let me check. I am sorry but telnet failed. Is there something else I could try?

*Yeah, can you try this other command – try pinging to the same IP address. [6]*

Sorry but I can't use this command because the ping command is firewalled.

*Can I just ask if you made any changes to the configuration? [6]*

No, I don't think so.

## 11.1 Dealing with a customer's problem: a ten-step solution (cont.)

Now let's analyse what the helpdesk operator does at each stage (indicated by the numbers in square brackets).

| | | |
|---|---|---|
| 1 | ANNOUNCE NAME OF YOUR COMPANY AND YOUR NAME. | Given that the caller may have called before, it is important to mention your name then callers may feel more relaxed to know that they are dealing with someone they already have a relationship with and who may be familiar with the type of problem they usually have. |
| 2 | OFFER HELP | You could also use a simpler phrase e.g. *How can I help?* |
| 3 | FIND CUSTOMER ON THE SYSTEM | By finding the customer's details you can see what problems, if any, this customer has had before. If you have an efficient system you could have explanations in English ready in response to this customer's problems (see 10.6). |
| 4 | REASSURE CUSTOMER | Instead of immediately saying *What is the problem?* you can adopt a relaxed manner. The caller may be agitated, so the idea is to try to calm the caller down. |
| 5 | REPEAT BACK WHAT CUSTOMER HAS SAID | This helps you to check that you have understood, and to give the customer an opportunity to clarify if necessary. |
| 6 | TRY TO ESTABLISH WHAT COULD HAVE CAUSED THE PROBLEM | Typically with software / hardware problems, you will need the customer to send the logfiles, but before doing so, it may help to investigate possible sources of the problem. |

The operator would then probably continue as follows with four final steps:

    7) for software related problems only: ask customer to send log files

    8) give customer a ticket number (i.e. a reference number that can be used to trace the issue / problem)

    9) reassure customer that the problem will be dealt with

    10) give customer a deadline within which the problem will probably be solved

## 11.2 What to say while the customer is explaining the problem

It is crucial that you do not remain silent while a customer is explaining their problem to you. In the conversation below Andi Gore (in *italics*) is a helpdesk operator for a company that produces engines. Note how she:

1. interjects (interrupts) frequently to show that she is listening carefully to what Fabrice (the customer) is saying
2. empathizes with the customer (see the two starred * phrases)
3. occasionally repeats back what Fabrice is saying – firstly to check her understanding and secondly to give Fabrice an opportunity to clarify

The combination of these three factors helps to reassure the customer that he / she is in good hands.

>*Andi Gore.*
>
>Hi Andi, this is Fabrice from …
>
>*Hi Fabrice how can I help you?*
>
>We're having a problem with the new engine.
>
>*Okay. Sorry to hear that *.*
>
>When we try to turn it on it makes a clicking noise but the engine does not crank.
>
>*Huh, huh alright.*
>
>So we checked the battery, but that's fully charged.
>
>*Right.*
>
>So we then checked the fuel.
>
>*You checked the fuel levels?*
>
>Yes, we did but there is no problem there either.
>
>*I see. And you're using the correct fuel, right?*
>
>Yes, of course.
>
>*OK. I am really sorry about this Fabrice but I am sure we can sort the problem in no time *. Can you give me the engine number?*
>
>Just a moment let me see… ah, it is 127p4
>
>*127p4 perfect. OK Fabrice, here's what I am going to do …*

## 11.3 Using questions to identify the problem

Before you take any action to solve a problem, you first need to be 100% sure that you know what the problem is. After you have let the customer explain their problem, you will probably need to follow up with a series of questions. In the questions below, the question forms are in *italics*.

> OK, let's try and figure out what's going on here. First of all, where *are you* calling from?
>
> And *can you* give me your name again please?
>
> And the product you are using is an X123, right? *
>
> And which version of the X123 *is it*?
>
> And you are seeing a red light in the dashboard, *is that* correct?
>
> And *what is* the error message that you are getting?
>
> *Have you tried* pressing the 'reset' button?
>
> *Are you sitting* in front of the dashboard now?
>
> OK, *could you* switch on the PQS. *Does a message come up* saying something like "status disconnected"?

Remember that when you formulate a question you should put the auxiliary first (e.g. *can, have, is*) and then the subject. But in any case, your priority should be in helping the customer, if you make a few grammatical mistakes with the question forms this should not interfere with the successful outcome of the phone call.

Note how in the starred * sentence the operator does not use a question form, but just says *right* at the end of the question. This is one of the standard ways of asking for confirmation.

## 11.4 Interrupting and repeating back what the customer tells you

To ensure that you identify the customer's problem, at each stage of the explanation, repeat back what they have said. You can repeat back in several ways:

1. by repeating the exact words
2. by paraphrasing what the customer has said
3. by interpreting what the customer has said and suggesting what the consequences are

For instance, if the client says: *The screen goes blank*. You could say:

> So the screen goes blank.
>
> So there is nothing on the screen.
>
> So you can't see anything and so obviously you can't navigate any more.

The advantages of this technique are that it enables:

- you to check that you have understood
- you to have a few seconds to think of the solution
- the customer to check that you have understood, and if necessary to revise or repeat his / her explanation.

Interrupting or asking for clarification has additional benefits:

- it makes the conversation more interactive and forces you to listen more carefully to what the client is saying. If you use the technique of repeating back what the customer has said, this implies that you have to concentrate on listening to what the customer says
- it shows the customer that you really want to solve their problem, i.e. that you are making a great effort to identify what the problem is

Below is an example of a conversation that illustrates how to interrupt and clarify. In this case both the caller and the helpdesk operator try to clarify what each other has said. The helpdesk operator's role is in *italics*.

> Hello, this is A from X. I am calling you because I have a problem with the FX console.
>
> *What version is it?*
>
> It is version 13.1.
>
> *Did you say thirteen point one or thirty point one?*
>
> 13.1

## 11.4 Interrupting and repeating back what the customer tells you (cont.)

*OK. What exactly is the problem you are having. Is it open, or are you having problems opening it?*

It's open but the add-in is not connected and I can't use it.

*You said that the add-in is not connected, but can you see some errors in the bottom of the window?*

Some errors. Sorry, what am I supposed to see?

*Can you see a red bar or a yellow bar in the bottom of the window with an explanation of the error?*

No, there don't seem to be any red or yellow bars.

*OK, no bars. So can you see a downloading bar in progress anywhere on the screens?*

No, I can't see any downloading bar in any of my screens.

*Are you sure, can you just check again?*

Sorry, yes you are right, there is a downloading bar in the third screen.

*Right OK. So if there's a downloading bar then your system is connecting to our system.*

So you are saying it is connecting?

*Yes, I think so. When did you press the connecting button?*

About 20 minutes ago.

*Twenty minutes ago. OK. We need to check the logs to see what the problem is. Could you email me the logs? I need you to send all the logs that you find in the logs directory.*

Sorry, where did you say they are?

*In the logs directory.*

In the what directory, sorry?

*Logs directory.*

Oh, logs. Sorry the line is bad. OK. Just a moment. I'll check. OK I have found them.

*OK, so now can you …?*

## 11.5 Suggesting possible causes and solutions: expressing certainty through adverbs and modal verbs

At some point in a conversation with your customer, he or she is likely to ask the question: *Do you have any idea what the problem might be?*

The following answers are in decreasing order of certainty going from 100 % certainty to about 50–40 %.

It's definitely due to …

It must be due to …

I am sure it is a result of …

It is highly likely that …

Very probably the cause is …

Possibly it is …

I would imagine it is …

It could be / might be due to …

Note that *can, should* and *have* to are not normally used in such circumstances.

~~It can be due to …~~

~~It should be the result of …~~

~~It has to be caused by …~~

## 11.6 Giving instructions to the customer

When you give instructions to the customer, use the imperative (i.e. the infinitive form of the verb without *to*). Using the imperative is not considered impolite. In any case, if you are worried that you might sound impolite, you can precede the imperative with *please*. Here are some examples, the imperative forms are in *italics*.

*Try* pressing the reset button.

*Download* the file from the customer area of our website.

Please *send* me the logfiles. [Alternatively: Can you / Could you please send me the logfiles].

*Feel* free to call me back if you have any further questions.

Note that *must, should* and *have* to are not normally used in such circumstances:

~~You should press the reset button.~~

~~You must download the file from the customer area of our website.~~

~~You have to send me the logfiles.~~

## 11.7 Telling the customer what you need from them and what the next step will be

Before you conclude the call you need to tell the customer:

- what if anything you need the customer to do
- what you are going to do next
- how soon you estimate the problem will be resolved
- that you are doing everything possible to sort out the problem

For example:

*What I am going to need from you, David, is the gateway logs ...*

*As soon as you can get those logs over, we can take a look at them. And it might be something really quick that we can see straight away from the log files.*

*You can send them via email and put my name in the subject line.*

*Otherwise you can upload the files directly onto our website in your customer area.*

*If you decide to upload the files directly onto our website, then let me know as soon as they are up, and I will pull them down. And hopefully we can figure it out as soon as possible.*

*I do understand that this is a major concern for you. We will try and sort it out as quickly as possible.*

*Don't worry, it sounds like a fairly routine issue.*

*I estimate that the problem will be solved in the next two hours. I hope that is OK for you.*

*I can't guarantee that it will be solved before tonight, but certainly early tomorrow morning.*

## 11.8 Showing the customer that you care

Below is a dialog between someone in support (in normal script) and a customer (in italics). Note how the helpdesk person is really concerned about the customer's situation, is following what they are saying, and gives the customer confidence that the problem will be solved quickly.

ABC. This is Andi.

*Hi Andi, this is Fabrice from PQR Frankfurt.*

Hi Fabrice how can I help you?

*We're having a problem with our Bloomberg gateway.*

Okay ...

*Last night we moved the gateway to a new environment, to a solaris box, it used to be on windows.*

Huh, huh alright ...

*Last night we moved everything and copied over the* negotiation.init file, *and all the other files.*

Right ...

*And this morning when the gateway started, we noticed that it has altered the negotiation. init file, and there is a line missing.*

A line missing -you mean the line has been deleted?

*Yes, we know at least one line, but maybe there are others ...*

Is this line at the beginning or at the end of the file?

*In the middle, it's a product group*

What version of the gateway is it?

*Just a moment let me see... ah, it is 127p4*

127p4 perfect. OK Fabrice, here's what I am going to need. Please could you send me the gateway logs, the psh logs and both versions of the negotiation.init files? I mean the original file and the one that has been altered.

*OK.*

Perfect – if you could send those via email. Just put my name in the subject line.

*OK, can I have a ticket number please?*

Sure, just give me a moment, let me pull it up for you... okay, the issue number is 24680.

*Thanks*

No problem Fabrice, I'll be waiting for those files. As soon as you send them over we'll try and get to the bottom of this.

## 11.9 Follow up with an email

When a customer has a problem that needs urgently resolving, they appreciate getting updates. A quick and easy way to achieve this is a simple email such as:

> *I just wanted to update you regarding your issue. Our technicians are currently looking at it and have promised to get back to me later this morning. As soon as I hear something I will let you know.*
>
> *In any case, rest assured that we are doing all we can to resolve the situation.*
>
> *Thank you for your patience.*

## Helpdesk – dealing with customers: Summary

- When you receive a call: 1) announce your name and the company's name, 2) say *How can I help?* or similar, 3) find the customer on your system to see if the problem has arisen before, 4) reassure the customer that you will be able to help, 5) repeat back what the customer says to check your understanding, 6) try find out what caused the problem, 7) give the customer a ticket number, 8) reassure the customer again that their problem will be dealt with, 9) tell the customer what you need them to do, 10) tell the customer when the problem will be solved.

- While the customer is explaining their problem, frequently ask questions (make sure you practice the question form in English), and empathize appropriately.

- Repeat back what the customer says to check your understanding and to reassure them that you really do understand their problem.

- Learn strategies for checking understanding.

- Use the imperative to give instructions to the customer.

- Show the customer that you are truly concerned about their situation and that you will help them solve their problem as soon as possible.

- Follow up the call with an email.

# 12 CALLING A HELPDESK

## 12.1 Facilitating a smooth service from the helpdesk operator

If you have read the previous two chapters you will realise that working on a helpdesk is not always an easy task. If you are a customer and you want your problem resolved quickly and effectively, always try to provide the following information:

- who you are and your company name
- what the product or service is; if you are talking about software then say how the application is configured and the machine where the application is running
- say what you were trying to do when you encountered the problem
- the result of your attempt
- the consequence for you if the problem is not resolved rapidly

Have a clear idea of what you want to say. If your ideas are not clear to you, they certainly won't be clear to your listener. This will also avoid you having to change track in mid-sentence, which might cause your listener considerable difficulty in following you.

Use short sentences, but remember that brevity doesn't always equal clarity.

Don't mention anything that is not strictly relevant as this is likely to confuse the helpdesk operator: he / she may not be able to distinguish what is relevant from what is not.

As you are giving your explanation, use the following phrases to check that the operator understands what you are saying:

> Are you with me?
>
> Is that clear?
>
> Does that make sense?
>
> Have you got that?

## 12.1 Facilitating a smooth service from the helpdesk operator (cont.)

Then when you listen to the operator's solutions and suggestions, ensure you understand what he / she has said by:

- repeating what the operator has said using other words
- asking for confirmation
- making occasional short summaries

## 12.2 How to interact with a helpdesk operator who has very poor English

If you are a non-native speaker then you know the difficulties that you sometimes encounter when speaking English. Remember these difficulties when you are speaking to another non-native who has very poor English skills. It is generally a good idea to:

- spell out names and numbers – repeat them several times
- stress the most important points clearly – repeat them several times.
- repeat / summarise / paraphrase using different words
- offer to write a summary via email

It is also helpful to check that the helpdesk operator has understood by never speaking for more than 30–45 seconds at a time and asking lots of questions:

*Is that clear?*

*Is there anything you didn't understand?*

*Do you need me to repeat that?*

*Am I speaking too quickly?*

*Have you got any questions?*

*Maybe it would be a good idea for you to tell me what you've understood so far.*

## Calling a helpdesk: Summary

- Inform the helpdesk operator: 1) who you are, 2) what the product / service is that is causing you problems, 3) what you were trying to do when you encountered the problem, 4) what steps you took and what the result was, 5) the consequence for you if the problem is not resolved rapidly.
- Use short sentences, but above all be clear.
- Only mention things that are strictly relevant.
- Constantly check what the operator has just told you.
- Ask if the operator has understood you – get them to summarize your problem (this will enable you to really check if they have understood or not).

# 13 PARTICIPATING IN AUDIO AND VIDEO CONFERENCE CALLS

Note: This chapter deals with how to be a <u>participant</u> in a conference call, video call or audio call. To learn how to be a <u>moderator / chairperson</u>, see Chapter 6 in the companion volume *Meetings, Negotiations and Socializing*. To learn how to conduct a training session via a video or audio conference see Chapters 11 and 12 in the companion volume *Presentations, Demos and Training Sessions*.

If you want to learn more about how to participate in a meeting (audio, video or face-to-face) see Chapter 5 in the companion volume *Meetings, Negotiations and Socializing*.

## 13.1 Audio conference calls

In a typical audio conference, participants:

- are in different locations
- cannot see each other, but they can see the same file on their monitor
- are in a very passive role and may find it very hard to concentrate on listening for more than 5–10 minutes at a time
- can send messages to the presenter / trainer and vice versa

Because you cannot see each other, you cannot:

- see how many participants are present
- establish any immediate relationship (e.g. there is no handshaking, no opportunities for chatting by the coffee machine)
- see / gauge people's reactions, i.e. you are not helped by being able to interpret facial reactions or negative body language; you cannot see if someone is nodding their head in agreement with what you are saying
- cannot connect with the others on a human level and you risk being an anonymous voice

This all means that you need to prepare for the call in order to make it as effective as possible.

## 13.2  Preparing for a conference call

If you are a participant, you will vastly improve the success of the conference call if you prepare for it in advance. Find out as much as you can about the meeting before you go, such as:

- the purpose of the meeting
- the topic
- who will be present (nationality, position in company, age)
- why you have been told / invited to participate
- what you will be expected to say

Then:

- decide exactly what you plan to say, then note down any key words and phrases in English that you might need.
- prepare a script of anything particularly important that you want / need to say. Practise reading your script aloud. Then modify it to make it more concise and convincing.
- think about the kinds of questions the moderator might ask you, and prepare answers (the following subsections contain examples of such questions).
- try to predict what people are likely to say. Write down some key phrases that will help you to agree with or counter what they might say. Are the participants likely to agree with what you are going to say? If not, think of ways in English to counteract their objections.

You also need to check if there is any documentation that you need to have in advance, such as the agenda and any files that participants will be expected to share.

In audio conference calls you cannot see the other participants. It will help you if you can at least 'picture' them. So if you have not met them before, try to find photographs of them: these will also indicate whether the person is male or female (you may not be able to understand this from their name).

## 13.3 Introducing yourself

You will need to know how the call functions and the difficulties involved. Typically, you will be called by an automatic operator who will ask you to 'hold the line' until all the participants are online. Alternatively, you may hear a beep when you join. At this point you should introduce yourself:

> Hi, this is Miranda. From the Oslo office.

The moderator (i.e. the chairperson managing the call) may also ask you to say something more about yourself. This will help the other participants tune into your voice. Try to speak slowly and clearly. A typical question the moderator may ask is:

> Miranda, glad you could be here. Can you tell us a bit more about yourself?

You can reply:

> Hi, so I am Miranda. I work in the Oslo office, although I am actually from the Czech Republic. I have been working for the company for nearly six months in the sales division.

Other typical questions that the moderator or trainer might ask you include:

> Can you tell me what your role is inside the company?
>
> Do you have any knowledge of the topic of today's demo already?

While people are joining the call, the moderator might ask questions such as the following. Before the call, think of how you would answer such questions:

> So Praveen, what's the weather like in Bangalore?
>
> Olga, how did the conference go?
>
> Here it is pouring with rain, what's it like with you?
>
> Milos, what time is it with you?
>
> Karthik, how was your holiday?
>
> Yohannes, how are things going in Ethiopia?

If you are late in joining the call, just say:

> Hi, sorry I am late. This is Miranda. From the Oslo office.

## 13.4 Dealing with technical and documentation problems

If you are having problems hearing the other participants you can say:

> This is Otto. I can't hear what you're saying – there's a high-pitched noise.
>
> Is someone on speaker phone because everything is echoing?
>
> Sorry, but I can't hear anything.

In a video conference another technical problem that you might encounter is with the video quality. You may need to say:

> Sorry but I can't see the slides very well. Could you make them a bit bigger?
>
> Sorry, but could everyone move a bit closer together as I can't see some of you.

Note that if the video quality is poor it may seem that your remote interlocutor is avoiding eye contact with you, but obviously this may not be the case.

Conference calls often involve looking at documents. Your moderator may ask you:

> Did you all get the files I sent you last night?
>
> Do you all have a copy of the agenda?
>
> Have you all got the presentation open at slide 1?
>
> Do you all have the documents in front of you?

If you are not sure what is being referred to you can say:

> Sorry what presentation are you talking about?
>
> Sorry, but I am not sure I received the document.
>
> Sorry, what slide are we supposed to be looking at now?
>
> Sorry, where can I find the doc?

## 13.5 Checking for clarifications during the call

If you are a non-native speaker and most of the other participants are native speakers then it is a good idea to say something like:

> English is not my first language, so it would be great if you could all speak very slowly and clearly.
>
> Speaking on behalf of the non native speakers, I would really appreciate it if you could all speak very slowly and clearly.

If you have problems understanding, using the chat facility to send messages can help resolve many difficulties. If you need to receive or make a clarification via chat, you can say:

> Sorry, I am not too clear about what Praveen said. Could you write it down for me?
>
> Can we just stop a second, I just need to make a note of what Clara was saying.
>
> Would it be OK to pause for a second and just use the chat? I am having difficulty in following this part of the discussion.

Given the fact that there are multiple participants, it is easy to get 'lost'. In any case, it is generally a good idea to interrupt.

> Sorry, I am not sure who is talking.
>
> This is Olga again. I'm sorry but it's hard to understand two people talking at once.
>
> Sorry, but the line isn't great at my end, could you all speak more slowly?
>
> Sorry, what slide are we up to?
>
> Sorry, what page are we on now?
>
> Sorry, I am not sure which figure you are talking about.

## 13.6 Ending the call

Typical things that the moderator might say to signal that the conference call is over include:

I think we've covered everything, so let's finish here.

Right, I think that about finishes it.

This is a good point to end the meeting.

Has anyone else got anything they want to add?

So this indicates your last chance to clarify anything that you did not understand. You also need to find out what you are expected to do next. You can say:

Sorry, I am not clear about what you want me to do next.

Sorry, but will you be sending minutes of the meetings? Or a summary?

## 13.7 Skype calls

Skype and similar systems can be used for videoconferencing. Bear in mind that:

- whether you will be able to use the video option may depend on the number of participants and how good the internet cable is
- sound quality may vary considerably from one participant to another – again this may depend on the line, but also on their PC, and how they are positioned in front of their PC

Given the above two points, it generally makes sense to have a sound / video quality control check before you begin the actual meeting. Typical phrases you may need are:

> Vishna, your voice isn't very loud, could you turn the volume up or sit nearer the microphone.
>
> Neervena, I can't see you very clearly – can you see me?
>
> I think we might be able to improve the sound quality if we turn the video off.
>
> OK, given that we have the video off, could I ask each of you to announce who you are before you say something [this will only apply when there are many participants who don't know each other].

A major advantage of Skype is that you can send written messages (see 14.7) to each other while you are speaking. So you can exploit this option if you are having difficulty understanding someone's English or when you need to clarify something that you are saying. You can say:

> Sorry, I am having some trouble understanding. Do you think you could just type the name of the product / website / document?
>
> Sorry, I am having difficulty saying the word. I am just going to type it for you.

## Participating in audio and video conferences: Summary

- An audio call requires specific planning in order to think of strategies to compensate for the fact that you cannot see the other participants.
- Before the call, find out about the purpose, topic, who will be there, why you are there, what you will be expected to say. Then write down and practise the key sentences that you will need to say during the call.
- Predict what questions you will be asked, and prepare answers.
- Introduce yourself slowly and clearly: name, office.
- Don't be afraid to say that there are sound problems, or that your English is not as good as the other participants.
- Constantly check that you have understood by asking questions or making mini summaries (provided of course that there are not too many people on the call).
- Use messaging systems to ask for / make clarifications.

# 14 WHAT TO DO AND SAY IF YOU DON'T UNDERSTAND

## 14.1 Foreign language skills of native English speakers

One reason why you may have difficulty understanding native English speakers is that they often have little or no idea that they are difficult to understand. This is because many native speakers only speak English. They have no experience of what it is like to have to listen to a language that is not their own. An incredible one third of the citizens of the USA feel that it is not too important or not important at all to speak a second language. The number of children and students studying languages in the UK has dropped considerably, and less money is being invested in language research.

Moreover, when they are feeling tired, some native speakers may be reluctant to converse with non native speakers: they believe that the effort required is much greater than when talking to another native speaker.

So what can you learn from the above information?

- If you do not understand a native speaker when he / she is talking, the problem will only be partially yours. For effective communication both speakers are equally responsible.

- You may need to make native speakers aware of the difficulties you have in understanding them – you need to 'educate' them to learn techniques that will help you to understand them better (e.g. enunciating more clearly, speaking more slowly, not using slang)

So to make it immediately clear to the native speaker that you need him or her to:

- speak slowly and clearly
- make frequent mini summaries for you
- be prepared for many interruptions for clarification on your part

## 14.1 Foreign language skills of native English Speakers (cont.)

You could say:

*Please could you speak really slowly and clearly, as my English listening skills are not very good. Also, I may need to ask for clarifications throughout the phone call.*

Even if the native speaker acknowledges your difficulties, they may forget these difficulties within two or three minutes when they become absorbed by what they are saying. So you can remind them to speak more slowly:

*I am sorry, but please could you speak more slowly.*

## 14.2 Ignoring words and expressions that you don't understand

In the dialog below some parts of what the speakers say have been deleted. These deletions are indicated with a line. Note that the line can represent from 8 up to 20 words. Does the conversation make sense in any case? Can you imagine what was said in the deleted text?

Hi Anna, it's Adrian here.

*Hi Adrian, how are things?*

Fine thanks. Listen [1] _____ I can't do the meeting at your office as planned.

*Oh.*

Yeah [2] _____ they need me here.

*I see, [3] _____*

OK, but what we could do instead is to do it via videoconference, but I certainly wouldn't be able to start before three.

*That would be fine by me, [4] _____*

Good.

Yeah but I'll need to confirm that with Gianni. [5] _____

*That's great. I hope I [6] _____*

Don't worry, that's fine.

Nearly 50% of the original dialog has been deleted. This illustrates that you don't need to understand every word in order to understand the main message.

Now look at the phrases that were originally in the six spaces above. Are you familiar with the expressions in italics?

1. I'm afraid I've got to *call off* tomorrow's meeting, or rather

2. the thing is there's been an *outbreak* of flu in the office and

3. *what a pain*, we were hoping to *wrap things up* by tomorrow evening.

4. that's only an hour later than we'd originally planned and should give us all the time we need.

5. Tell you what, I'll just *give him a buzz* and then get back to you via email.

6. haven't *messed up* your arrangements too much.

There are three main conclusions that can be drawn from this dialog, and specifically from the six phrases above.

## 14.2 Ignoring words and expressions that you don't understand (cont.)

Firstly, several of the expressions in italics are phrasal verbs (*call off, mess up, wrap up*), which are commonly used by native speakers. When seen out of context it is almost impossible to work out (i.e. identify, understand) their exact meaning. However the above dialog illustrates that the surrounding context should help you to understand what they mean, and even if you don't hear them at all, the overall meaning of what is being said should be clear.

Secondly, words are often used in combination, where in reality only one of the words is necessary. For example, in the case of *outbreak of flu*, you don't really need to understand *outbreak* (which means a scenario where a lot of people have caught an illness or virus). And even if you didn't understand the whole phrase (*outbreak of flu*), it would not matter because the main point that the speaker is making is that he cannot attend the meeting – the reason why he cannot attend is incidental.

Thirdly, even if you don't understand the meaning of an expression (e.g. *what a pain*), a clue to the meaning will probably be given by the speaker's intonation. In this case, do you think the speaker is expressing something positive or negative? It is clear from the context that *what a pain* means the opposite to *that's great news*.

Obviously there will be cases where not understanding expressions could be a problem. However, there are strategies for dealing with such cases. These strategies are outlined in the rest of this chapter.

## 14.3 Don't say 'repeat please'

Here is an extract from a telephone conversation between a native speaker (in italics) and a non-native speaker (in normal script), which highlights the dangers of simply saying 'repeat please'

*Hi, could I speak to Jake please.*

Sorry he's at lunch, do you want to leave a message?

*Yeah could you tell him the meeting's been put off till next Thursday.*

Repeat please.

*Could you tell him the meeting's been put off till next Thursday.*

Sorry, could you speak more loudly please.

*The meeting's been put off till next Thursday.*

Repeat please, more slowly.

*The ... meeting ... has ... been ... put ... off ... till next Thursday.*

Sorry I need to ask a colleague for help, I don't understand.

If you say 'repeat please' your interlocutor does not know whether you understood a few words or absolutely nothing. Consequently, the interlocutor repeats the whole phrase, probably in exactly the same way and at the same speed as he / she did before. The non-native speaker in the dialog above then hopes to resolve the situation by asking the native speaker to speak more loudly – again the interlocutor repeats almost the same sequence of words but this time louder (but probably still too rapidly). At this point, both parties are frustrated. After an attempt at speaking more slowly, the conversation then ends.

Probably what the non-native speaker did not understand was the expression 'put off' (*put off* is a phrasal verb frequently used by native speakers, but offering no indication of its meaning to a non-native).

The dialog below highlights a more effective approach.

*Hi, could I speak to Jake please.*

Sorry he's not at his desk, do you want to leave a message?

*Yeah could you tell him the meeting's been put off till next Thursday.*

Sorry, 'put off'?

*Postponed. The meeting will now be on Thursday.*

I see, the meeting has been postponed until Thursday?

*Yes, that's right*

## 14.3 Don't say 'repeat please' (cont.)

In this case, the non-native immediately identifies the words that are causing problems, and says 'Sorry, put off'. The word *sorry* immediately indicates to the native speaker that the non-native speaker has not understood something. So, the native speaker understands that 'put off' is the problem and so thinks of a synonym (*postponed*).

Of course, the native speaker could interpret that the non-native is just confirming rather than trying to understand. The native speaker may then say 'Yes, put off'. If this happens, then you can say:

*Sorry, what does* put off *mean?*

## 14.4 Choose the quickest and easiest way to indicate exactly what you don't understand

In the case above, the non-native could try to understand the native speaker in many different ways:

*Sorry, put off?* (1)

*Sorry, you mean postponed?* (2)

*Sorry, what does put off mean?* (3)

*Sorry, do you mean postponed?* (4)

*Sorry, I don't understand [the meaning of] put off.* (5)

*Sorry, do you mean that the meeting has changed day and is now on Thursday?* (6)

The alternatives above are listed in the order in which they would be quickest and easiest to say.

When you are having difficulty in understanding, you are likely to be more anxious and your English language skills will probably suffer. So begin the clarification process with the simplest grammatical form.

In Form 1 you simply repeat the same words that you think your interlocutor said. If you didn't understand the individual words, you can imitate the general sound that you think you heard.

In Form 2 you suggest a meaning for the word you did not understand.

Forms 3, 4 and 5 require you to formulate a question or a negative (by using an auxiliary verb), and thus require more mental effort to remember the construction. Form 6 is very explicit, but does require a great deal of effort to formulate.

So my suggestion is that you start with an easy form (1 or 2). Then if this does not achieve the required result, then use a more complex form.

In any case, don't worry about the grammar, just focus on being able to understand.

## 14.5 More examples of asking for clarification and making comments

In the examples below the part in *italics* is what was said by your interlocutor that you did not understand.

Points a) and b) are possible requests for clarification that you could ask. Point c) is a comment you could make if, on the other hand, you did understand what was said.

*My car broke down.*

    a) Your what broke down?

    b) You car did what?

    c) Oh dear, so how did you get here?

*On Saturday I went to the cinema.*

    a) You went where?

    b) Sorry, where did you go?

    c) Oh yes, what did you see?

*I saw Stefan.*

    a) You saw who.

    b) Who did you see sorry?

    c) So how is he? I haven't seen him for ages.

*This weekend I drafted the report.*

    a) You wrote the report?

    b) What does drafted mean?

    c) Did it take long?

*We need it next week.*

    a) Next week?

    b) For when, sorry?

    c) So you're saying you need it for next week? That sounds fine.

## 14.5 More examples of asking for clarification and making comments (cont.)

*I found a wallet on the street.*

    a) A what, sorry?

    b) Sorry, where did you find it?

    c) Did it have any ID in it?

*Kalinda needs some help to consolidate the data.*

    a) Sorry, could you say the last part again?

    b) She needs help to what sorry?

    c) You mean she wants me to help her? No problem.

## 14.6 Distinguish between similar sounding words

Some words sound very similar to each other and are frequently confused, even by native speakers. Below are some examples of how to clarify certain pairs of words.

| WORDS | POSSIBLE MISUNDERSTANDING | CLARIFICATION |
|---|---|---|
| Tuesday vs Thursday | We have scheduled the meeting for Tuesday. | That's Tuesday the sixth right? |
| 13 vs 30 | We need thirty copies. | That's thirty, three zero, right? |
| can vs can't | I can come to the meeting | So you are saying that you <u>are</u> able to come to the meeting? |
|  | I can't come to the presentation | So you mean that you are <u>not</u> able to attend the presentation? So you mean that you <u>cannot</u> attend? |

In the first example, the secret is to combine the day of the week with its related date. This means that your interlocutor has two opportunities to verify that you have understood correctly. If you have misunderstood, your interlocutor can then say: *No, Thursday the eighth.*

The confusion in the second example happens with numbers from 13 to 19 and 30, 40, 50 etc. Using the correct stress can help: thir<u>een</u> vs <u>thir</u>ty. However, particularly on the telephone, this subtle difference in pronunciation may not be heard. So the secret is to say the number as a word (e.g. *one hundred and fourteen*) and then to divide it up into digits (*that's one one four*). If you have misunderstood, your interlocutor can then say: *No, thirteen, one three.*

In the third example, the problem is increased if *can* is followed by a verb that begins with the letter T. Thus understanding the difference between *I can tell you* and *I can't tell you* is very difficult. There are also significant differences between the way native speakers pronounce the word *can't* – for example, in my pronunciation *can't* rhymes with *aren't*, but for others the vowel sound of 'a' is the same as in *and*. The solution is to replace *can* and *can't* with the verb *to be able to*. You also need to stress *are* in the affirmative version, and *not* in the negative version, as illustrated in the table. If you have misunderstood, your interlocutor can then say: *No, I <u>am</u> able to come* or *no, I am <u>not</u> able to come* (alternatively *I <u>cannot</u> attend*).

## 14.7 Use instant messaging systems

You can massively improve your understanding of a phone call, if you combine the oral / listening element, with a written element. If there are some words, expressions or particular details you don't understand, then ask the caller to write you a message (and you can do the same for them). You can say:

> Sorry, I think it would be easier for me if you wrote down the address rather than dictating it to me. Could you send me a message?
>
> Is it OK if I just write that down for you and message it to you?

## 14.8 If you really can't understand, learn a way to close the call

There will be occasions when you simply cannot understand. Rather than panicking and putting the phone down with no explanation you can say:

- *I am afraid the line is really bad. I will try calling back later.* You can then prepare yourself better for the next time you call, or alternatively ask a colleague to make the phone call for you

- *I think it might be better just to send an email. I will do this as soon as possible.* You can summarize in an email what you think you have understood, and then ask for clarifications

## 14.9 Reasons why understanding a native English speaker can be difficult

When we read a text, the punctuation (commas, full stops, capital letters etc) help us to move within a sentence and from one sentence to the next. Brackets, for example, show us that something is an example or of secondary importance. Punctuation also helps us to skim through the text without having to read or understand every single word. We don't really need to read every single word as we can recognise certain patterns and we can often predict what the next phrase is going to say.

A similar process takes place when we listen to someone speaking our native language. We don't need to concentrate on every word they say. Much of what people say is redundant – we all use a lot of filler expressions and incomplete phrases. However, we can automatically understand from their intonation and the stress they put on certain words, when they are beginning and ending a phrase, and what elements are important within the phrase and what elements are in 'brackets'.

Unfortunately, although we can usually quite easily transfer our reading skills from our own language into another, we cannot transfer our listening skills – particularly in the case of the English language. English often sounds like one long flow of sounds and it is difficult to hear the separations between one word and the next (usually because there aren't any!).

However spoken English does follow some regular patterns, and if you can recognize these patterns it may help you to understand more of what you hear and enable you to understand the general meaning rather than trying to focus on individual words and then getting lost!

In the written language pauses are denoted by commas (,) and periods (.), in the spoken language pauses within a speaker's speech may be due to the speaker's 'online planning' of what he or she is going to say next. Because we are thinking online, we often begin phrases and project our intonation in a particular way, but then we may abandon what we are saying – even in the middle of a word. We then either go back to 'repair' what we have just said (which is often indicated by a slight change in intonation), or we just leave the phrase incomplete. Thus, unlike the written language, which generally has some logical sequence, the spoken language often seems to follow no logical track and is therefore more difficult to understand. However by recognizing the intonation we can get a clearer idea of the 'direction' in which the speech is going.

The basic units of written language are clauses, sentences and paragraphs. Spoken language tends to have much shorter units, which may not have subjects or verbs and would be considered incomplete in the written language. Individual streams of speech are sometimes known as 'idea units' as they contain generally one idea or one new piece of

## 14.9 Reasons why understanding a native English speaker can be difficult (cont.)

information. They are also called 'intonation units' or 'tone units' because each unit is part of one intonation flow.

In the dialog below the first speaker is in normal script and the second speaker in italics.

The tone units are divided up by slashes (/). A tone unit typically begins with a stronger (louder, clearer) voice and ends with a weaker (softer and less distinct) voice.

As a listener, you just need to focus on the key words (in bold in the dialog) within each tone unit. If you can understand these key words then you should be able to get the gist (i.e. the overall meaning) of what your interlocutor is saying.

**ABC** / This is **Andi**. / How can I **help** you?

*Hi, i need to **speak** to Andrea **Marchesi**.*

Andrea is **not here** today. / He's in our **London** office / but until **9 am** London time / calls are **routed** here to **Pisa**. / May I ask **who's calling**?

*This is **David***

Okay, what I can do **David**, / is take your **number** and a **message** / and I can try and have **Andrea** call you **back**. / Would that be **OK?** / Or, if you like, / you can wait till after 9 am London time / and **try** him again.

*No, that's alright. / Could you ask him to **call** me?*

Sure, can you **spell** your **last** name for me?

*Yes it's Marchesi.*

Another reason you may have difficulty understanding spoken English is that some very common words may not be pronounced in the same way as you have learned them. In fact, there are around 50 frequently used words that have two pronunciations. One is used when they are stressed (which is probably how you have learned them) and another when they are unstressed.

Examples of these words are:

articles: *the, a, an*

prepositions: *for, to, at*

modals and auxiliaries: *can, could, must, would, have, has*

pronouns: *you, them, he, his, him*

## 14.9 Reasons why understanding a native English speaker can be difficult (cont.)

conjunctions: *and, but, that*

to be: *are, be, been, was, were*

The above examples are generally used in their weak forms. The strong forms are only used:

i) when a modal or auxiliary verb appears at the end of the sentence:

Have you been here before? (weak or strong)

Yes, I have. (strong only)

ii) when auxiliaries are used in their full form with 'not':

I haven't seen this before. (strong)

iii) for emphasis

Have you given it to him yet? (weak)

I gave it to her not him. (strong)

Why haven't you done it yet?

I have done it. (strong)

Spoken English is essentially driven by stress which means that unstressed words that are located between stressed ones are often squashed together or swallowed completely. This means that they are often difficult to decipher. The same may happen when someone is talking fast.

A lot of people when talking make noises between words or phrases to give themselves time to think. Typical noises are *er, erm,* and *um.* The problem is that you may think these are separate meaningful words or that they are part of other words. It is thus important to recognise and be aware that these are only noises.

A frequent characteristic of the spoken language is when words have their initial or final sound clipped (cut). A typical example of initial clipping is *because* which you may have seen written as *'cause, 'cos* or *'cuz*, particularly in music lyrics. End clipping is more common, particularly the final *g* in the ing form, e.g. *goin', lovin'*, again typical of singers but also very common in the normal spoken language. The final *t* in many words is often not pronounced when it is immediately followed by another word, e.g. *las(t) year*.

## 14.9 Reasons why understanding a native English speaker can be difficult (cont.)

Finally, another reason why people may be difficult to understand is because they tend to use a lot of vague language or they fill their speech with redundant words and expressions. Here are some examples:

if you know what I mean

like

and stuff like that

or something

and things

something like that

## Improving your understanding of what the caller is saying: Summary

- Communication is two-way. If you do not understand, part of the responsibility is with your interlocutor. Make native speakers aware of the difficulties that you have in understanding them, and tell them how they can help you. Don't be embarrassed to do this.

- You don't need to understand every word to understand the key message. So relax, and just try to understand what you can. Remember that you can always use email.

- Learn the quickest way to indicate to the interlocutor that you haven't understood something that they have just said.

- Distinguish between similar sounding words, and check understanding (e.g. say *the meeting is on Thursday the sixth,* instead of *the meeting is on Thursday*).

- Use all the technologies possible to make the phone call easier for you, e.g. use the messaging systems of Skype.

- As a last resort, ask the caller to email you.

- Learn to focus on the key words and don't be distracted by the words and phrases that you do not understand.

# 15 USING THE WEB AND TV TO IMPROVE YOUR LISTENING SKILLS

## 15.1 Set yourself a realistic objective

Many people rely on English lessons to improve their English without considering the fact that you can learn a considerable amount outside the classroom.

Given that a major difficulty for most people is understanding native English speakers, it makes sense to take every opportunity you can to listen to English on whatever platform you have. This chapter gives advice on what to watch and how often.

Be realistic about what you want to achieve. Just as you can't lose 10 kg in weight in one weekend, you can't learn English in two days.

Here are some ideas.

1. watch the news in English once a day or one presentation on TED (see 15.6)
2. watch two episodes of a 20-minute TV series or one episode of a 50-minute episode

But be patient. The first 10-20 times you watch you will only understand 10%. Then you will gradually manage to reach about 50%, and then hopefully even more.

Only watch things that you would have watched anyway in your own language. You are only going to learn if you are motivated and enjoy yourself.

Expose yourself to lots of different accents.

## 15.2 The news

Listening to the international news on TV or on the radio is good practice because you are probably already aware of some of the stories and you will thus be able to follow them much better. Watching news that is all local to one country is much more difficult.

## 15.3 YouTube

There are hundreds of thousands of short videos that you can watch on YouTube. The vast majority have no subtitles. You can find videos on:

- English pronunciation and listening skills for specific nationalities
- Lessons on giving presentations and communicating in general
- Seminars and talks by top business people and experts
- Presentations from conferences

## 15.4 Dragon's Den

This TV program, which you can find on YouTube, was originally a Japanese 'reality show' in which contestants present their inventions, products and services in front of venture capitalists (known as dragons – these are real people with real money to invest).

There are four English language versions: British, Canadian, Irish and US (under the name Shark Attack). The Canadian version is fun and fast moving, but the British version is perhaps the easiest to understand, though rather serious. Dragon's Den makes great viewing because you will:

- learn useful business terminology and related phrases
- learn how to pitch (present in a very short period, one or two minutes) an idea
- see some incredibly good (and bad) inventions

Dragon's Den has the same advantage as a TV series (see 15.5), in that the main 'characters' – i.e. the venture capitalist – are the same throughout each series (though they sometimes change from series to series). This means that you will have a chance to get used to the dragons' voices and manner of speaking, thus you will tend to understand more and more, the more you watch.

## 15.5 TV series

The main advantages of TV drama series are that they:

- can often be compulsive viewing, so you are really motivated to watch the next episode
- are shorter than movies so this makes finding time to watch them much easier
- show characters who keep re-appearing so that you get tuned in to the voices, also these characters tend to have particular phrases that they say repeatedly
- go on for years, so you have a constant source of entertainment

## 15.6 TED

Ted.com is packed with fascinating talks given by experts in various fields from around the world. You will learn not only English, but also interesting facts. You can:

- choose the topic you are interested in by using their internal search engine - the main topics are technology, entertainment, design, business, science, culture, arts, and global issues
- choose the speaker
- choose the most watched talks, the most recent talks, the most talked about talks
- choose the length of the talk depending on the time you have available for watching - they vary from a minimum of around two minutes up to a maximum of around 20 minutes
- read a transcript of the talk either in English or your own language - you can do this before you watch to give you a clear idea of the topic, and also while you are watching. The transcript is interactive in the sense that you can click on words within the transcript and be automatically taken to that same point in the video
- use the subtitles - there are English subtitles for all the talks, and for the very popular talks there are often subtitles in many other languages
- download the talk, and play it on other media
- read comments made by people who have watched the talk and contribute to a discussion on the topic

Note that the existence or not of subtitles and translations into various languages depends on how recently the talk was posted (if it is within the last few months it may not have either of these features) and how popular the talk is.

If you are really serious about improving your listening, then another option is to copy and paste the transcript and invent your own listening exercises by deleting random words and then while listening you can try to fill in the gaps. It helps if you delete the words a few days before you then listen, otherwise you might remember what you deleted.

## 15.7 Movies

Movies are often hard to understand because:

- the plot is totally new
- the voices are all new
- the film tends to last at least 90 minutes, which requires intense concentration

Nevertheless, watching is movies is fun. So

- choose films you have already seen in your native language, so then you do not have to worry about following the plot
- check out on YouTube to hear what the actors sound like, and try to find extracts from the film to judge whether you are likely to enjoy it and understand it
- consider watching it over several days

The easiest movies to understand are:

- science fiction - full of technical words that you may be familiar with, and there is little humor (humor tends to be quite difficult to understand)
- documentaries - the narrator tends to speak clearly and from a script, so even though you can't see the lips moving you will still be able to understand
- historical

The most difficult are ones that contain non-standard English, ones with lots of slang, and ones with a lot of humor and thus full of word plays, for instance:

- ones containing dialects
- comedies
- thrillers and crime stories

You might also consider watching old movies. Ones that precede the 1970s tend to be a little easier to understand as the actors tended to enunciate the words more clearly. Also, the plot is slower and easier to follow.

## 15.8 Subtitles

There are no rules for the use of subtitles. The main problem is that if you use subtitles you will probably read the subtitles rather than listen. However, try to watch some parts of the video with subtitles and some without; or watch a part with subtitles and then re-watch the same part without subtitles.

If you do opt for subtitles, I suggest that you use English subtitles - select 'English for the hard of hearing'.

If there is a part that you simply don't understand then switch on the subtitles from your language.

It is also fun to watch films in your own language with English subtitles!

## 15.9 Songs

If you have any favorite English-speaking bands or singers, then try listening to their songs while reading the lyrics. They may contain a lot of slang but the ear-training that you will get will be very useful.

## 15.10 Audio books and podcasts

You can buy audio books and download podcasts on a huge variety of topics. If you put them on your iPod you can then listen to them while you are traveling. The BBC provides a lot of downloadable materials (bbc.com).

## 15.11 Other websites worth checking out

You may find the following websites useful for improving your English listening skills.

http://esl.about.com/od/englishlistening/a/intro_podcasts.htm

http://iteslj.org/links/ESL/Listening/Podcasts/

http://www.bbc.co.uk/programmes/b00s9mms

http://www.betteratenglish.com/

http://www.eslpod.com/website/index_new.html#

http://www.podcastsinenglish.com/index.shtml

## Using the web and TV to improve your listening skills: Summary

- Be realistic about how much listening / watching you will find time to do in the space of a week. Set yourself feasible objectives.
- Watch a mix of programs: news, ted.com, TV series, movies etc.
- Prefer short programs (e.g. news, ted, short TV series) in order not to lose motivation.
- Discover ted.com and Dragon's Den - they are not only really interesting, but you will find your English listening improves without you making too much effort.
- Use subtitles with care - do not rely on them, but at the same time use them when you are in difficulty or are tired.
- Check out the vast range of listening materials on the web specifically designed for non-native speakers.

# 16 PRONUNCIATION: WORD AND SENTENCE STRESS

## 16.1 Investigate free software that will help you to improve your pronunciation

The major difficulty people have when using the telephone is to understand what the other person is saying. This could be because the line is bad, you are nervous, or the other person is speaking to fast. But often it is because their pronunciation of a particular word does not match your own pronunciation of the same word. If you think the pronunciation of the term 'reset' (as in the 'reset' button on a modem) is /reizet/ then you are unlikely to understand the term if it is pronounced /riset/.

Just because your colleagues in your own country can understand your English, it certainly does not mean that people from other countries will understand you.

When learning languages we often meet a word for the first time when we are reading (rather than listening). As we are reading we assign a pronunciation to any new words we meet without actually knowing whether that pronunciation is the correct pronunciation or not.

It is a good idea to write a list of words that you think you may need for a specific occasion. Then you can use one of the following to enable you to hear the correct stress and vowel sound for each word.

GOOGLE TRANSLATE

Google Translate is designed to translate a text from one language into another.

However, you can also use its text-to-speech option, by clicking on the sound icon. Thus you can type in a text and hear it being read by a male voice in a reasonably slow and clear way.

You can also hear how a text in English might sound if it was read aloud by someone of your own nationality. For example, if you are a Spanish speaker, you can insert a text in English but tell Google that the text is Spanish. A Spanish voice will then read your text. Besides being fun, this should show you how you might sound when you yourself speak in English!

## 16.1 Investigate free software that will help you to improve your pronunciation (cont.)

NATURAL READERS

This free software (http://www.naturalreaders.com) enables you to 'read aloud any text' and you can also 'adjust the speed' and change the voice. Being able to adjust the speed, e.g. to slow it down considerably, is very useful. Most read aloud systems have voices that speak too fast for you to really hear how, for example, a multisyllable word is stressed.

HOWJSAY.COM AND OTHER PRONOUNCING DICTIONARIES

With this website you can type in (submit) a specific word and then hear the pronunciation of that word. The site even gives alternative pronunciations of the same word. For example *innovative* can be pronounced with the stress on the first (in*no*vative) or third syllable (inno*va*tive). Although this is a British site, it sometimes gives you differences between US and GB pronunciation, so that you will learn that *tomato* is pronounced one way in Britain and another in the United States, but it does not do this for all the differences between these two varieties of English.

If you just want the US pronunciation you can try: www.merriam-webster.com/dictionary/pronounce

Another good online dictionary of both US and GB pronunciations is: http://dictionary.cambridge.org/help/phonetics.html

ODD CAST

This free online application (http://www.oddcast.com/home/demos/tts/tts_example.php) is basically just for fun. You can type in a word or text and hear various voices (male, female, many different nationalities) pronouncing the text for you.

ADOBE 'READ ALOUD'

If you convert any document – a report, a presentation, a letter – into a pdf file, you can then use the 'read aloud' feature to 'hear' your document. Personally, with respect to Google Translate and Natural Reader, I find the voice slightly more mechanical and a lot faster. However, by the time you read this book Adobe might have produced a more natural sounding reader.

The rest of this chapter explains how to decide which syllable to stress in a word.

## 16.2 Two syllables: general rules

Most two-syllable nouns and adjectives have the stress on the first syllable. In fact, the vast majority of British surnames and place names have the stress on the first syllable. Examples: *Thatcher, Newton, London, Bristol.* This is not the case for many US names as these have been more influenced by American Indian, Spanish and the languages of the settlers: *Los Angeles, New York.*

|  | SOME EXAMPLES | SOME EXCEPTIONS |
|---|---|---|
| NOUNS | China, country, effort, colleague, method, minute, person, program, statement | amount, canal, cartoon, cassette, effect, event, exam, guitar, hotel, Japan, police, success, technique |
| ADJECTIVES | common, English, famous, heavy, perfect, pretty, private, previous, recent, various | afraid, aware, Chinese, complete, concise, correct, polite, precise |

Most two-syllable verbs, conjunctions, prepositions, adverbs have the stress on the second syllable.

|  | SOME EXAMPLES | SOME EXCEPTIONS |
|---|---|---|
| VERBS | allow, accept, depend, forget, support | happen, listen, wonder, manage, mention |
| ADVERBS | ago, before, perhaps, towards, until | after, also, likewise, seldom |

## 16.3 Two syllables: same word (noun on first, verb on second)

Some words change stress depending on whether they are nouns (on first syllable) or verbs (on second syllable).

Examples: contact, exploit, increase, insert, object, present, progress, record, report, research, upset.

In the table below all the words are both verbs and nouns.

| FIRST SYLLABLE | | SECOND SYLLABLE | |
|---|---|---|---|
| answer | invoice | address | reply |
| access | measure | command | report |
| archive | promise | concern | request |
| comment | profit | control | respect |
| contact | question | correct | result |
| discount | schedule | debate | return |
| issue | survey | effect | support |

Unfortunately, some words are stressed differently by different people, e.g. *research* – some people say *research* and other *research*, irrespectively of whether it is a noun or verb. Also, the British say *detail*, and the Americans *detail*.

## 16.4 Compound nouns

A word that is made up of two words has the stress on the first syllable (e.g. *software*). Here are some examples:

> boyfriend, everyone, feedback, headline, highlight, income, input, interface, interview, layout, newspaper, outcome, overview, podcast, sidetrack, supermarket, switchboard, workshop

Exceptions: afternoon, understand

## 16.5 Three syllables

Most three syllable words (nouns, verbs and adjectives) that don't have a suffix (e.g. *un-, pre-),* have the stress on the first or second syllable. Only a few have the stress on the third syllable (e.g. expertise, introduce, Japanese, personnel).

| FIRST SYLLABLE | | SECOND SYLLABLE | |
|---|---|---|---|
| absolute | hierarchy | acceptance | embarrass |
| agency | industry | accompany | example |
| alias | influence | advantage | explicit |
| apparent | interested | assistant | financial |
| architect | interesting | component | ideal |
| article | modify | configure | important |
| atmosphere | monitor | consultant | objective |
| attitude | paragraph | convenient | percentage |
| bicycle | personal | determine | performance |
| company | prejudice | develop | strategic |
| conference | premises | dishonest | sufficient |
| confident | principle | | |
| consequence | quality | | |
| deficit | satellite | | |
| difficult | sufficient | | |
| excellent | triangle | | |

## 16.6 Multi-syllable words

Words ending with *-able, -ary, -ise, -ize, -yse, -ure* have the stress on the first syllable:

suitable, secretary, category, realize, analyze, organize, recognize, architecture, literature

Words ending with *–ate, -ical, -ity, -ment, -ology* have the stress on the third to last syllable:

graduate, immediate, separate, logical, reality, feasibility, management, development, government, environment, psychology

Some exceptions: equipment, fulfillment

Words ending with *-ial, - ic, -cian; - sion, -tion* have the stress on the penultimate syllable:

appearance, artificial, specific, expensive, politician, occasion, specialization

Most words ending in -ee have the stress on -ee:

attendee, employee, interviewee referee

An exception: committee

The majority of other multisyllable words have the stress on the second syllable (e.g. identify, particular, parameter, enthusiasm), but some on the third (e.g. fundamental, correspondence).

A number of commonly used multisyllable words are usually pronounced without certain syllables (i.e. the ones in italics below are not pronounced in normal speech):

ave*r*age, bus*i*ness, categ*o*ry, Cath*o*lic, comfo*r*table, diffe*r*ence, eve*n*ing, gene*r*al, inte*r*ested, inte*r*esting, labo*r*atory, lite*r*ature, med*i*cine, prefe*r*able, refe*r*ence, tempe*r*ature, veg*e*table, Wed*n*esday

## 16.7 Acronyms

Acronyms are pronounced in three ways:

1. with each individual letter pronounced separately and with equal stress on each letter, e.g. DVD, EU, UN, WWW

2. like a normal word, e.g. NATO, UNESCO, URL

3. like a normal word but with vowel sounds added, e.g. FTSE (pronounced *footsie*)

## 16.8 Sentence stress

Generally you should stress the word that carries the key information or that helps to distinguish one thing from something else. This means that normally we stress adjectives rather than their nouns:

*I am a* software *developer.*

You would only stress the noun if it is the noun that helps to differentiate between two things.

*I am a software* developer *not a software salesperson.*

Stress verbs rather than pronouns:

*I want to* show *you.*

Only stress the pronoun when you want to differentiate one group of things or people from another.

*I want to show* you *not* them.

Stress the main verb rather than an affirmative auxiliary, unless you want to give special emphasis.

*This has* happened *several times.*

*I can assure that this* has *happened several times.*

Stress the negative auxiliary rather than main verb, unless you are distinguishing between two verbs.

*This* hasn't *happened before.*

I haven't *spoken* to him but I have seen him.

# 17 EXAMPLE TELEPHONE DIALOGS

This chapter shows you various types of telephone calls made in a business context. You can use these scripts to:

- learn how a phone conversation is typically structured
- see typical phrases in context (for more on typical phrases see Chapter 18)
- practise reading aloud

In all the examples, the caller is in *italics* and the receiver in normal script.

The majority of these examples are transcripts of real calls made at the offices of companies where I have trained staff.

## 17.1 Switchboard operator: trying to connect someone

ABC Incorporated, good morning can I help you?

*This is George from the Madrid office, could you put me through to Marion please.*

Sorry, could you say your name again please.

*It's George from the Madrid office.*

Sorry, but the line is really bad today. I didn't catch your name, could you spell it for me please?

*Of course: G E O R G E*

Oh, OK, George. Could you hold the line please, I'll put you through to Marion. Pause Hello, sorry she is not at her desk at the moment, would you mind calling back later please?

*Yes of course. Thanks. Bye.*

## 17.2 Switchboard: Taking down someone's name and number

I'm afraid she's not at her desk at the moment, I think she's just gone out to lunch.

*Do you know what time she will back?*

Well she normally takes about half an hour. Shall I get her to call you as soon as she comes back?

*Actually could you possibly give me her mobile number. It's quite urgent.*

I am really sorry but I am not authorized to.

*Is there any chance of you ringing her for me and asking her to call me back?*

Certainly, I can do that for you. Could I have your name please.

*Yes it's Penny Dalgarno.*

Sorry what was your surname again?

*Penny Dalgarno. d a l g a r n o*

OK Penny I'll ring her straight away. Goodbye.

## 17.3 Switchboard: checking understanding

I am sorry could you speak up a bit, I can't hear you very well.

*Sorry, is that better? I was saying that I would like to speak to the sales manager.*

To the what manager, sorry?

*The sales manager.*

OK, sorry she's not in at the moment would you like me to take a message?

*Yes, could you ask her to ring me. It's David Milsom.*

Sorry I didn't catch your last name.

*Milsom, that's M-I-L-S-O-M. And I'm calling from meta4.*

Could you spell that for me please?

*Yes it's m e t a and then the figure four, have you got that?*

The figure for what sorry?

*Sorry what I mean is that it is meta in letters and then the number 4, I know it's a strange name isn't it? And my number is 020 8347 1254, and I am calling with regard to distribution problems in the London area.*

Sorry I didn't get what you said in the middle, what kind of problems?

*Distribution problems. Look would it be easier if I just called back later?*


## 17.4 Switchboard operator: chit chat

ABC Incorporated

*Hi Gabry, it's Marina, how are you?*

Fine thanks and you?

*Fine , so what's the weather like in Madrid?*

It has been raining all week.

*Oh dear. That's too bad. In Pisa it has been sunny since Monday. Listen, the reason I'm calling is because I have some problems with ... blah blah blah*

## 17.5 Switchboard: dealing with an employee who rings in sick

Good morning, ABC Incorporated. This is Andi. How can I help you?

*Hi, this is Rafael.*

Hi Rafael how are you?

*Not so good. I don't think I am going to be in the office today.*

Ah OK. Don't worry, I'll send an email out to the New York office. Would you like me to send it to any other email address?

*Yes, could you send to qa@abc as well?*

Ok, no problem, I hope you feel better soon.

*Thanks.*

You're welcome, bye.

*Bye.*

## 17.6 Switchboard: giving out a phone number

ABC Incorporated this is Andi (how can I help you)

*Hi, is Brent there?*

No, I'm afraid he's traveling today, can I ask who's calling?

*Ah, well could you give me his mobile number?*

I'm afraid I don't have that, but if you give me your name, I can try and have him call you back.

*It's Jane, from the London office.*

Ah Jane, sorry, I didn't realise it was you. Brent is going to be in the London office this afternoon. Was it something important?

*Yes it's about a delivery schedule.*

Alright, let me see if I can get you his mobile number, hold on a second ...

*Thanks.*

OK, it's 348.123.4567.

*So that's 348.132.4567*

No, 23 pause 4567

*OK thanks.*

No problem, bye.

*Bye.*

## 17.7 Switchboard: dealing with a caller whose request you cannot fulfill

HI can I speak to Susan?

*HI the London office is closed at the moment, and your call has been redirected here. So I can take a message if you like. If you could give me your name and phone number I'll get her to call you. Or alternatively you could ring back after 9.*

Could I have her direct number please?

*I'm afraid we're not authorized to give out direct numbers. So the best thing is if you could give me your phone number and I'll ask her to call you.*


## 17.8 Leaving a message with the switchboard

Hi, could I speak to Jake please.

*Sorry he's not at his desk, do you want to leave a message?*

Yeah could you tell him the meeting's been put off till next Thursday.

*That's Thursday the seventh right.*

That's the one, and that there should be thirteen people coming from the New York office.

*Is that one three or three zero?*

One three, thirteen. And could you also tell him that we can't test out the software beforehand.

*Sorry you can test it or you can't?*

We can't.

*Sorry the line's really bad, do you mean that you are or are not able to test the software?*

No, we will not be able to test it.

*Can I just check I've got everything?*

Sure, thanks.

*Meeting on Thursday the seventh, thirteen people coming, no testing beforehand.*

That's it. Sorry I wasted all your time I could have just sent him an email.

*No problem. I've just remembered you haven't even given me your name!*

You're right, it's …

## 17.9 Arranging a meeting

Could I speak to Damian Shurst please.

*Speaking.*

Hi Damian, it's Emma Tomson.

*Oh Emma, I didn't recognise your voice, how are you doing?*

Fine thanks and you? How's your new job going?

*Well, it only started last week, but it's going very well thanks.*

Listen Damian, we need to fix a date for the meeting. What about Thursday at nine?

*That sounds fine. See you on Thursday then.*

Thanks Damian, be seeing you.


## 17.10 Changing the time of a meeting

Hi Anna, it's Adrian here.

*Hi Adrian, how are things?*

Fine thanks. Listen I'm afraid I've got to call off tomorrow's meeting, or at least I can't do the meeting at your office as planned.

*Oh.*

Yeah the thing is there's been an outbreak of flu in the office and they need me here.

*I see, what a pain, we were hoping to wrap things up by tomorrow evening.*

OK, but what we could do instead is to do it via videoconference, but I certainly wouldn't be able to start before three.

*That would be fine by me, that's only an hour later than we'd originally planned and should give us all the time we need.*

Good.

*Yeah but I'll need to confirm that with Gianni. Tell you what, I'll just give him a buzz and then get back to you via email.*

That's great. I hope I haven't messed up your arrangements too much.

*Don't worry, that's fine.*

## 17.11 Sales division: Dealing with a customer inquiry

Chandra Karamadni speaking.

*Good morning this is Damo Suzuki from ABC. I'm calling to see if it's possible to have the new version of XXX on a trial basis.*

Yes of course we could arrange that for you. You can have a one-month trial period.

*Would we get the complete product or just a demo version?*

It's the complete product. After the month's trial the product automatically stops working.

*Is it possible to have additional features, I mean functions that are made to measure for our company?*

Well you would obviously have to specify your requirements and then we would see whether it's possible to make the modifications or additions that you requested.

*What if we find some bugs?*

If you discover any bugs then a new version will be sent.

*If after the month's trial period we decide that we would like to purchase the product, what is the procedure?*

You simply send us a fax informing / telling us that you agree to our offer. In any case at the beginning of the trial period we will send you the licence agreement and the conditions of sale.

*How much is the basic cost?*

The basic configuration is ten thousand euros.

*And is that a one-off licence?*

No, that's the annual licence fee.

*And is there a discount if we decide to buy several copies?*

Yes, we can give you a discount, but it will obviously depend on the numbers involved.

*Would you be willing to give me some names of companies who are already using your products?*

Well, I'd have to check with them first, but yes it should be possible.

*Going back to the cost. Does the ten thousand euros include subsequent upgrades, or are they charged separately?*

Yes, it includes any upgrades.

## 17.11 Sales division: Dealing with a customer inquiry (cont.)

*OK. Well I think that's about everything.*

*Well if you have any further questions then please give me a call.*

*OK. Thanks very much for your help.*

*You're welcome. Goodbye.*

*Bye.*

## 17.12 Making an enquiry about a company

*Sakawi Inc. This is Eriko from sales.*

This is Naomi King from ABC.

*Good morning Ms King, how can I help you?*

I'd like to know something about what Sakawi does.

*Well, we have ten years experience in developing products both in Japan and in Europe for the electronic money markets,. For examples, MTS ... We not only develop products but also projects, so we work specifically for companies in creating dedicated products for them. Our clients are both banks and financial institutes.*

We work both in and outside Asia and we have offices in London too.

*Sakawi also has offices in London. So perhaps it might be more convenient for you to meet at our offices in London. You can contact Mr Ryosei Akibi in Japan. Could you send me an email specifying exactly what it is you want to know.*

OK, I'll do that right away.

*Shall I give you my email address? It's e dot morita at sakawi dot jp. 'jp' as in Japan. E as in Eriko, which is my first name. When I get your mail, I'll send you more information about the company and our field of activity. I'll also send you the addresses of our two websites. At the Sakawi website you can find first level information about our products, and at Sakawi dot UK you can find details of a particular product called ViewKeeper.*

What's your role in Sakawi?

*I work in the marketing department. And could you just tell me something about your company. What are your main activities?*

Blah blah blah.

*Well, it's been very nice talking to you Ms King. I look forward to receiving your email.*

Thank you. Goodbye.

*Goodbye.*

## 17.13   Outlining / Solving technical problems 1

Good morning. How can I help you?

*I am having a problem with the UST gateway.*

Right.

*I'm doing a test, but when I run it, I can't connect to the market.*

Which type of connection are you talking about?

*Post trade.*

OK, I'll check (for you) and I'll let you know when I finish.

*Thanks very much.*

Not at all, bye bye.

*Bye.*


## 17.14   Outlining / Solving technical problems 2

*Listen, I wonder if you could help me with a problem I am having. I started the router on the Madrid platform,*

Right.

*But when I connect the X, the connection remains blank. Do you have any idea what the problem might be?*

Could you check the IP address ...

*Right I'll do that immediately.*

And could you send me the log files please.

*Yes of course. I'll download the files on manta. I'll call you back in 10 minutes. Let me know if you have any problems.*

Great. Thanks very much.

## 17.15 Outlining / Solving technical problems 3

I'm calling about the problem with the AQ that one of our customers is having.

*OK I think I already know which problem you are talking about. This behaviour is due to a bug that I think was introduced some time ago but which we have now nearly fixed.*

OK thanks for your help. If I find any problems, I will call you again if that's OK with you.

*Sure, no problem. Feel free to call whenever you want.*

## 17.16 Extracts from a conference call

Kim: Hi this is Kim, is anyone else there?

Lee: Hi Kim it's Lee. How are things?

Kim: Fine thanks. Is everyone picking up all right?

Thierre: Yeah this is Thierre. I can hear you fine.

Helmut: This is Helmut. I can't hear what you're saying – there's a high-pitched noise going on.

Kim: Is that any better?

Helmut: That's fine now.

Kim: OK, Cosmo's here now – let's begin.

Kim: Can I just remind you all to say your name when you speak. We have a couple of people on the call who are not native speakers and if this call is to be successful, we need the native speakers to speak as clearly as possible. *Pause* The goal of this call is to discuss the key items of x, y and z. OK Thierre do you want to start?

Thierre: No you go first.

Cosmo: Sorry, who's that speaking?

Thierre: Sorry, it's me Thierre.

Lee: Can I just say that …

Kim: I'm sorry but it's hard to hear two people at once

Thierre: This is Thierre again.

Jessica: Hi, Jessica here . Sorry I am late.

Kim: Hi Jessica, could you wait a sec. Then I'll recap everything for you.

Kim: Cosmo, what are your thoughts on that?

Cosmo: I have nothing to add actually.

Thierre: OK, we're moving on to slide 3 now.

Lee: Sorry, what slide are we up to?

# 18 USEFUL PHRASES

In this chapter are lists of useful phrases that you can use at various stages of a telephone call. In each case, several alternatives are given. I suggest that you learn the alternatives that you think will be the most useful and that are also the easiest to remember. However, it you should also be familiar with the other phrases: even if you don't use them yourself, you need to recognize them in case the caller says them to you.

Note: sentences *in italics* indicate what the caller has said. The sentence immediately after the phrase in italics, is your response to what the caller has said. Italics are also used where a word should be substituted for the word in italics (e.g. name of company, your name, place).

## 18.1 Switchboard: saying / establishing who is calling

### Receiving a call: Saying who you are

*Name of company*. Good morning. *Your name* speaking. How can I help you?

### When you recognize who's calling

Oh good morning David, what can I do for you?

### Asking for name / company

Could I ask the name of your company?

Could you repeat the name of your company?

Who shall I say is calling?

Who's calling please?

### Ways to get them to repeat their name / name of company / person desired

Sorry the line's bad. Could you tell me your name again and where you're calling from?

Sorry could you speak more slowly – the line is really bad.

Sorry, who did you want to speak to?

Sorry, who did you say you wanted to speak to?

Sorry, I didn't catch who you wanted to speak to.

## You're not the right person to deal with the call

I'm sorry but I'm not the right person, I'll put you through to someone who can help you.

Sorry, I'm new here - I'll just get someone who can help you.

I'm sorry but I haven't been working here for very long.

Actually, I think you need to speak to the secretary.

I'll just put you through to the secretary, who should be able to help you.

Sorry but I don't think I am the right person to help you. You could try ringing this number and ask for Luigi.

I'm sorry but I can't answer that question.

I'm not the right person, but I'll make sure someone from the office rings you back.

## Saying that you cannot contact the desired person at the present moment

I'm sorry but this switchboard is not connected to the building where Mr X is. So I'm afraid you'll have to dial again on this number …

I'm sorry but I'm alone at the switchboard at the moment and there's no one else around, so I'm afraid I can't give her your message straightaway. But as soon as someone comes back, I'll give them your message.

## Attempting to connect to the person desired

Hello Mr Thomas, I'll see if I can put you through.

Hold the line please, I'll see if she's back from lunch.

I'll put you on hold.

I'll just put you on hold.

Would you like to hold?

Just a moment … ringing for you.

Sorry to have kept you waiting.

Sorry to keep you waiting Mr Thomas.

Sorry to have kept you waiting.

I can put you through to Mr Rossi now.

## Person desired not in the office

I'm sorry but the person you need is not available at the moment.

She's on the other line / in a meeting.

He's on holiday / vacation until next Tuesday.

He's taken a day off.

I'm afraid you've just missed him.

I'm sorry but she's not in her office.

He's held up at a meeting.

She's with a client.

She's just taking her lunch break.

## Saying when person desired will be available

She's not in the office yet / this morning – she should be here in the next hour.

She should be at her desk in within an hour.

She's at lunch and should be back in about five minutes.

He should be back in about half an hour.

He has an appointment this afternoon and won't be back until 5.

He'll be back at the beginning of next week.

## Offering help when person desired is not available

Hello, the line's engaged, shall I try again or would you like to call back later?

Can I put you through to her assistant?

Would you like to refer the matter to me and I'll see if I can put you through to someone who can help you?

May I take your name and number and get someone else to call you back?

When would be the best time to reach you?

Is there anything I can do to help?

Is there anyone else you'd like to speak to?

Can I take a message?

## Responding to the caller's requests

*Could you ask him to phone me?* Yes of course.

*Could you confirm the date of the meeting with my secretary?* OK. I'll do that.

*I'd like to speak to someone in sales.* OK, I'll just put you through.

*Is the marketing manager in?* No, I'm afraid he's just left the office.

**When you can't talk for some reason**

I'm sorry, but I'll have to call you back in five minutes.

Are you still there?

Sorry about that. I just had to open the door for someone.

**Dealing with a wrong number**

No, this is *name of company*. I think you must have got the wrong number.

What number did you dial?

No, our number is 181 not 182.

You must have dialed the wrong number.

Are you sure you've got the right number?

I think you must have been given the wrong number.

**Person requested does not work at your company**

I'm sorry there's no one here by that name. We don't have a Smithson here.

I'm afraid Helen White no longer works here. I can give you her new number if you like.

I'll just check it for you.

## 18.2 Calling: saying who you are and who you want to speak to

**Saying who you are**

Good morning this is *your name*. I'm calling from *name of company*.

Hi Louise, this is Virginia. Sorry to disturb you, but …

**Saying who you want to speak to (formal)**

Could I speak to *name* please? *Speaking*.

This is *name*, could I speak to *name* please?

I'd like to speak to someone about …

I'd like to speak to someone in production.

I wonder if I could speak to Mr Smithson.

Well, if Mr Smithson is not in, could I speak to his secretary?

Could you put me through to Andrea Thomas / the personnel department.

I would like to speak to Mr Smith from administration.

Could you give me extension 218 please?

### Saying who you want to speak to (informal)

Hi, this is Virginia. Could I speak to Louise?

Is Louise there?

Is Henri still around?

Is *name* in this morning?

Is *name* there please?

### Asking who you are speaking to

Who am I speaking to?

Sorry, who am I speaking to?

Is that you Jennifer?

Is that Louise? / Is that Louise speaking? *Yes, speaking. / Yes, it's me.*

Sorry, I didn't recognize your voice.

### Asking for someone whose name you don't remember

I've just been speaking to someone from your department about x, but I didn't catch their name / I forgot to ask their name.

I was speaking to a man / woman yesterday afternoon.

I was speaking to someone called James in the x department, but I don't have his last name or phone number, do you have any idea who it might be?

## 18.3  Calling: when person desired is not available

### Responding to switchboard operator's offers to help

*Would you like me to put you on hold?*

Yes, thank you. / No thanks I'll call back later.

*Would you like to refer the matter to me and I'll see if I can put you through to someone who can help you?*

Well actually I could just leave a message with you if that's OK.

Yes, if you could find me someone that would be great.

*May I take your name and number and get someone else to call you back?*

Yes, my name is … and she can get me on *phone number*.

*When would be the best time to reach you?*

At about three o'clock London time.

### Doublechecking that the person desired really is not available

Would you check / Can you just check if he's in another office please?

Are you sure he isn't there? Mr Rossi had arranged to call him at this time.

### Checking when person will be available

Can you tell me what time he'll be back?

When are you expecting her back?

When can I get hold of him?

Do you when she will be in?

When will he be back from lunch?

### Asking to be called back

Could he possibly call me back? My name's Rossi and my number is …

Could you ask him to call me?

Could you get him to call me?

Could he give me a call?

Could he possibly call me back as soon as he comes back as it's rather urgent?

Could you just tell her I called?

She'll know why I called.

Could you possibly email me the details that you've just given me?

Can I just check the number?

### Saying when you'll call back

OK, if he's in his office this afternoon, I'll call back at around five.

Right, I'll call again next week.

## 18.4 Initiating the call with the desired person

### Asking if this is a good time to call

Is this a good time to call?

Sorry, is this a good time?

Sorry to bother you but I just wanted to get back to you with ...

Have you got a couple of minutes?

### Asking when you can call back

*Actually I'm really busy at the moment.* So when would be a good time to call?

### Making chit chat

How are you? *Fine thanks. And you?*

How's it going? *I'm rather busy at the moment actually.*


## 18.5 Leaving a message

### Asking to leave a message

This is *name* calling from *company* in *place*. Could I leave a message for Hillary Binns please?

Can I leave a message?

Could I leave a message with someone from administration?

Do you think I could leave a message with his secretary?

*Name* asked me to call him / her back.

I just wanted to check / confirm / ask if ...

Could you ask him / her to call me back?

I'll be in the office today until ...

Would you mind asking him to call me on my mobile?

### Asking when someone will be available

Do you when Heinrich will be in?

When will Miroslav be in?

When will he be back from lunch?

Could you get him to call me?

Could you ask him to call me?

Can you tell him it's very urgent, there is a problem with the ….

## Spelling out names and numbers

Could you ask her to ring me back on 02 878 705 (oh two / eight seven eight / seven oh five)

I'm sorry but I gave you the wrong number. It's two <u>one</u> six, not two <u>three</u> six.

Shall I spell that for you?

I'll spell that again for you.

No, there are <u>two</u> Ls in Hillary, not <u>one</u>.

That's seventeen - one seven.

No, that's Winker with an 'i' not an 'a'.

Yes, that's right.

## Confirming that the operator has taken your message down correctly

*I'll just read that back to you.* OK.

*So that's 0181 980 4187.* Yes, that's right / Perfect.

… *four one six seven.* No, four one <u>seven</u> <u>six</u>.

… *r-o-s-i.* No, it's double s: r-o-s-s-i.

Have you got that?

Could you just read that back to me?

## Suggesting that email is the best option

Would you like me to email that to you?

It's a bit complicated isn't it? Shall I email it instead?

I'm sorry, but would you mind emailing that to me, as I'm not sure if I've got it all correctly.

Can I confirm that by email?

## Asking for and giving email / website address

Could you give me your / his email address please?

Her address is: anna_southern at virgilio dot it. That's anna A-N-N-A underscore southern at …

Is that one word or two?

His address is adrianwallwork at virgilio.it. That's adrianwallwork all one word with no dots.

The website is: metaphor dot it forward slash marketing [metaphor.it/marketing]

**Leaving a voicemail**

This is *name* from *name* of company.

It's 12.30 in the morning our time.

I've been trying to get hold of you all day. But I keep getting your machine.

Do you think you could ring me back when you get in?

This is where I'll be at until 12.00, …

After that you can ring me at the office.

Hear from you soon. Bye.

## 18.6    Taking a message

**Asking to take a message**

Would you like to leave a message?

Can I take a message?

Would you like me to give her a message?

Can I ask what it's about?

**Asking for details**

Let me just check that I've got that right.

Who was it that you spoke with yesterday?

Shall I tell *name* to call you back?

Does *name* have your number?

Can I take your name and number?

When is the best time for him / her to call you?

What is your daytime contact number?

Are you always contactable on this number?

**Checking**

Can you spell that please?

Could I just spell that back to you?

Can I read that back to you?

So the number is 0208 388 6070?

So that's …

**Concluding call**

OK, I'll make sure she gets your message.

I'll tell him / her you called

I'll let him / her know you called.

I'll refer that to him and I'll get him to call you back.


## 18.7  Problems with understanding

**Asking for repetition**

I'm sorry I didn't quite catch that.

Sorry, what did you say?

Would you mind repeating that please?

Could you say that again please?

What did you say your name was?

Your name was?

Can you repeat that last part please.

Can you go over the bit about …

I'm not that clear about …

I'm sorry, I didn't catch that.

I'm sorry, what did you say?

A 'what' sorry?

Can you spell that for me?

I'm still not sure what you mean by 'x'?

I'm sorry I still don't understand.

**Problems with voice and speed**

Do you think you could speak up a little, please?

Could you speak a little more slowly please?

## Using technical problems as an excuse for not understanding

I'm sorry but the line's bad (I can hardly hear you),

I'm so sorry we got cut off.

The line's very faint.

I'm sorry, I didn't hear that, I'm afraid you're breaking up.

Sorry, the line is really bad, can you speak up please?

We might get cut off in a minute, if we do, I will call you straight back.

I'm sorry, I'm going to have to call you back, I can't hear you.

Sorry, I just missed the last part of what you said.

It's a terrible line I can hardly hear you.

## Suggesting calling back

I can hear you - but not very well, let me call you back.

I'm afraid there's something wrong with the line, can I call you back?

I think I'd better call you back the line is terrible.

Do you think you could call me back?

The line is terrible. Would you mind calling me back?


## 18.8  Checking and clarifying

### Checking that the other person has understood

Would you like me to repeat my name and number?

Have you got that?

Would you mind repeating that back to me?

Is that clear?

Am I making myself clear?

Does that seem to make sense?

Does that sound OK to you?

Is there anything you're not quite clear about?

Would you like me to go over anything again?

Are you OK with that?

**Confirming that you've understood**

Yeah, that's fine.

Yes, that sounds fine.

Yes that makes sense.

Yes, I'm clear about that.

I've got that.

Yes, I've got all that.

**Checking that you've understood**

So do you mean that …?

Are you saying that …?

Sorry, I still don't see what you mean.

I'm still not sure what you mean.

**Confirming that other person has understood**

Yes, that's right.

Exactly.

**Checking that you've taken everything down**

Have I got everything?

Is that everything?

**Apologizing for misunderstandings**

Sorry, I didn't mean to …

Sorry, I thought you meant …

I meant …

I didn't mean to offend.

Sorry I obviously didn't make myself clear.


## 18.9 Calling someone you already know: giving background to your call / updating

**Giving reason for your call**

I'm calling (you) to …

I'm calling about …

The reason I'm ringing is to find out if you have any news about the order I faxed you yesterday.

Anyway the reason I'm calling is …

I will try and keep it short.

## Asking for an update

How's it going?

Is everything working OK?

What's the latest on … ?

Have you got any news about … ?

Are there any problems?

Have you made the orders that I asked you to?

Is there anything else you need to tell me?

Have you had a chance to look at the order I emailed you a few hours ago?

I just wanted to call you to see where we're up to with …

I was wondering if you could do me a favor. Could you tell me …

I was just wondering if you had had time to check the …

Are there any problems?

How's it going with the project?

Have you made the orders that I asked you to?

Is everything working OK?

Is there anything else you need to tell me?

## Giving an update

I've looked into the …

I've had a word with …

I've examined your requirements …

## Telling someone how to proceed

Could you ring me back before 12.00 please.

I'd be grateful if you could give me an answer by this evening.

Can you get back to me first thing tomorrow?

Could you fax that to me.

Could you send me confirmation by email.

Then will you call me back and tell me …?

**Asking how someone will proceed**

Are you going to email them to me?

When can I expect your call?

Do you think you'll be able to get back to me before the end of the day?

**Suggesting how someone should proceed**

You could contact administration who should have the details. / Why don't you contact administration?

The best thing to do would be to contact administration as they …

One idea might be to contact …

You could always contact …

Why don't you check whether …?

**Informing someone how you will proceed**

OK I'll send them to you in a few minutes.

I'll get back to you before 6.00 tonight.

I'll be in touch later today.

I'll send you the information you required first thing tomorrow.

I'll put them in the post straightaway.

If you don't hear from me you can assume that everything is OK.

**Asking how you should proceed**

What would you like me to do with it?

What would you like me to do?

Please let me know what you'd like me to do.

Shall I … ?

Do you want me to … ?

**Moving on to a new subject**

While I'm at it …

While we're on the phone …

By the way …

## 18.10 Calling someone back

**When you call someone back**

This is *name* returning your call.

I believe you rang this morning.

I had a message to ring you.

Hi Pete, this is *name*. I believe you called earlier on.

I'm sorry I wasn't in when you called. I had just popped out to the bar.

**Calling someone back who you've just spoken to**

Sorry to bother you again but …

Hello, it's me again. I just wanted …

Hi, this is Cristiano again. I've talked to Tim and …

**When someone calls you back**

Thanks for getting back to me.

I called you because …

Actually, I've already managed to sort out the problem but thanks for calling back anyway.

**When someone calls you back and you had been about to call them**

I was just about to ring you.

Thanks, I was actually going to call you back.

Actually, I was just about to ring you back.


## 18.11 Requests / enquiries

**For information**

I wonder if you can help me.

Could you tell me whether .. ?

I was wondering if …

Do you happen to know if …?

I'm looking for …

I need some information about …

### Responding

I don't know offhand – I can easily look it up for you.

No but I'll try and find out for you.

I'll just check for you. What exactly do you need to know?

Sorry to keep you waiting. I've asked a colleague and was told that …

Thanks for holding. I think I've got the information you were looking for.

I'm sorry I couldn't be of more help.

### For someone to do something for you

Could you show it to Mr Rossi and ask him …

Could you check whether I sent you …

Could you possibly show this to Andrea.

Please could you ask them to sign the order.

Can you just make sure you have received all the other orders.

Can you just check that you received …

Would you mind sending it again?

### Responding

No, I haven't looked at it yet. I'll just go and get it. OK I've got it now.

No, no problem. / No, not at all. / Of course.

I'll send it straightaway.

OK I'll do it straightaway.

I'll do it as soon as he gets back from lunch.

I'll do it first thing tomorrow morning.

Listen, I'm just a bit busy at the moment. Could I ring you back in an hour?

### Thanking

Sorry to have troubled you.

Thanks very much.

That would be great. Thanks.

Brilliant. Cheers.

## 18.12  Cold calling (calling a company for the first time)

### Asking if this is a good time to call
Did I catch you at a good time?

Do you have a moment to speak?

Are you the right person to speak to about … ?

### Explaining where you got their number from
Your name was given to me by *name*, who thought you …

I got your number from …

### Outlining what your company does
We help IT companies like yours to achieve …

We are a marketing company that helps companies like yours with …

Our software allows computer-based organizations like yours to …

We can supply you with …

### Giving more details about your role in the company
I'm in charge of …

I am responsible for …

I'm the sales manager …

For the last few months I've been dealing with …

### Moving to the next stage
Can I make a note of your email address so I can send you that information?

Is it OK if I call back some time next week to see if you have any questions?

What is the best time to call you? Is this the best number to reach you at?

If you are interested in finding out more we could schedule a demo for you and your team.

So we've agreed that I will send you our brochures as well as the customer references you asked for. And you've said you will get in touch with your purchasing manager - when do you think you'll have a chance to speak to her?

### Concluding the call
Thanks so much for your time. It was very useful speaking with you.

## 18.13 Making a complaint, registering a problem, calling a helpdesk

### Making a complaint

I'd like to make a complaint.

I'm afraid I have a serious complaint to make.

### Making the same complaint again

[*Aggressive*] Listen, this is the third call I've made this morning.

[*Polite*] I wonder if you can help, I've been trying all morning to find someone who can …

I've already given my name and number … and no one has returned my call.

### Describing the problem

We've been experiencing problems with …

We seem to be having problems with …

It looks as if it might be the X which is causing the problem.

If we're not mistaken, it must be the X.

There seems to be a problem with …

I'm afraid there's a problem with …

It seems you forgot to attach …

### Saying what you have done so far to try and resolve the problem

We've tried to solve this issue ourselves by …

I consulted your manual and / but …

We've done everything we can from our end, but we are unable to …

### Underlining the urgency of the situation and giving the helpdesk a deadline

We need this problem solving …

   immediately.

   by the end of today.

   by the end of the week at the latest.

If it is not solved by today then we will have to …

The consequences for us are very serious.

Without x we are unable to do y.

I am sure you will appreciate that this requires your immediate attention.

### Reacting to solutions offered by the supplier

That sounds like it would be feasible.

That sounds reasonable.

That depends.

I don't think that would be possible.

Thanks for you time and effort.

I'll wait for your response / call.

### Ensuring that in the future you can speak to the same person again

If I have any more problems can I phone you directly?

Can you give me your direct line please?


## 18.14   Helpdesk: finding out about the problem

### Initial reaction upon hearing customer's problem

Sorry to hear about your problems with x, but hopefully we can get to the root of the problem so it will not happen again.

It sounds like it could be an x problem.

### Finding out how serious the problem is

How long has this been happening for?

How many users have the same problem?

Is this the first time it's happened?

Was it the first time it had happened?

Do you know if by any chance you changed something yesterday? I mean, that might explain why this has happened.

Is there any other information you can give me that you think might help us solve the problem?

### When you don't understand what they are talking about from a technical point of view

Sorry, would you mind if I get one of my colleagues to ring you back, because I am not sure that I can answer your question myself.

Sorry I'm not too familiar with this component. I think the best thing is for me to note down the problem and then get back to you by the end of the day. Would that be OK?

What you're telling me sounds a bit strange, I don't think we've ever encountered something like this. Are you absolutely sure that …?

**Gaining time**

I'm sorry, I can't say right now, I'll have to get back to you on that.

I'm not the best person to answer that, let me just put you on to my colleague.

I'll need to work it out and let you know.

Would you mind sending me an email briefly outlining the problem?

Hmm, good question.

Just a moment.

Ok, let's see what we can do, just hold the line for a moment.

I'll just confirm that with …

## 18.15  Helpdesk: dealing with a problem

**Giving instructions on what you need in order to solve the problem**

I need you to email me the name of the product, the product part number and your client ID.

I think what would be good is if you could send me a …

Do you think you could possibly send me a copy of the warranty?

First, could you send us your workspace that you were using when the problem occurred?

Second, we need you to send to us the libraries file, here's how to get it:

Lastly, do you remember which cells you were moving between when the error occurred, or any other information that could help us diagnose the problem?

**Diagnosing the problem**

We have examined the x that you sent, and it seems that …

Perhaps this is causing a problem with the …

The result is that …

**Making suggestions**

I wanted to ask if there was any possibility of ...

Would you be prepared to ...?

What if we ...?

## 18.16 Helpdesk: checking that you have both understood each other

**Getting caller them to repeat their question**

Sorry, I'm not really sure if I've understood what you need.

Sorry, do you think you could start again?

Sorry, what was your question again?

I'm really sorry but I didn't catch your question.

Sorry, I didn't understand the first part of your problem.

Could I just take down the question, then I'll be able to check for you.

**Confirming what you have understood**

Just so that I'm clear ...

Can I just check that I've understood your question correctly.

So, basically what you're saying is that you'd like to know how to do x.

I'm just going to check that I have understood everything you've said. Interrupt me if I've got anything wrong, thanks.

**When the caller doesn't understand you**

Sorry, my English isn't that great, I'll start again.

So does what I am saying seem to make sense?

Is that clear so far?

Are you clear about that so that we can move on to the next part?

So are you now clear about what this is and what it does?

**When you don't understand the caller / Clarifying**

Can you repeat that please?

Sorry, I didn't catch that.

Sorry, which department did you say?

Did you say seventeen or seventy?

When was the delivery date?

What was your question?

I just wanted to check …

I wanted to ask about …

Ok, so, just to confirm …

Let me just check that …

Let me just read that back to you …


## 18.17  Helpdesk: summarizing the problem, outlining a solution

**Summarizing the problem**

So what you're saying is …

Let me just check I've understood the situation.

In a nutshell then …

**Outlining a solution**

I'll look into the matter, and ring you back.

I'll get on to that problem immediately.

OK, let me tell you what we can do here.

Right, I think the first thing to do is to …

So what we're going to do is …

This is what I'm going to do.

**Apologizing**

I'm sorry if we've caused you any problems.

I'm very sorry but I'm unable to help you right at this moment but I have noted the problem and will get back to you as soon as possible.

Unfortunately I don't have an immediate solution for you. I will speak to someone more specialized in this area and get back to you as soon as possible.

Do you think you could write down what you've said in an email and then send it to …

Could you possibly email me what you've just said so that I can:
> ... be sure that I haven't misunderstood anything.
> ... then forward it to the person concerned.

**Saying when you will give them an answer**

Would it be OK with you if I got back to you later this morning?

Can I get back to you tomorrow with an answer?

**Responding to their thanks**

*Thanks very much for your help.*

You're welcome.

Not at all.

No problem.

Feel free to call me if you have any further problems.

Let's hope we've fixed it.

**Concluding the call**

I think that's all.

Anything else I can help you with?

Is that everything?

OK, then, have a great weekend.

Have a nice evening.

Great. Thanks for your call.

## 18.18 Being a good listener: reassuring and empathizing

**Showing that you are listening to the caller**

OK.

Huh, huh, alright.

Right.

**Empathizing**

Yes, I quite understand what you mean.

I can see exactly why you're not satisfied.

This must have caused you a lot of trouble.

I'm sorry about that.

Oh that's terrible.

Oh dear.

What a nuisance.

How annoying.

**Reassuring**

Well I think that now we're in a position to do something concrete about this.

It will only be a matter of hours before things are rolling smoothly again.

I'll get on to the matter immediately.

I'll make sure it gets sorted out straight away.

I'll personally make sure it doesn't happen again.

You should have the information by tomorrow morning.

I'll do everything I can.

That's going to be difficult but I'll certainly do everything I can.

**Giving a running commentary / Telling caller what you are doing**

The file's just opening now.

I'm just thinking that through.

I'm just looking for you now.

**Filling silences**

Just a minute.

Hold on please.

One moment please.

I'll be with you in a second.

Could you just give me a moment?

## 18.19 Apologizing

**Generic**

I'm really sorry.

I'm sorry about that.

**For not having done something**

I'm sorry about that. I'll get on to it straight away.

**Responding to apologies**

Oh that's alright.

Don't worry.

Not to worry.

These things happen.

No problem.

## 18.20  Thanking

**a Generic**

Thank you.

Thanks very much.

You've been really helpful.

Cheers.

Thank you very much for your help.

**b Responding to thanks**

Not at all.

You're welcome.

Don't mention it.

## 18.21  Arranging a meeting for yourself

**Giving reason for meeting**

I'd like to arrange a meeting to speak about …

Could we possibly meet up some time and discuss x?

**Suggesting time / place**

What time would suit you?

Would next Thursday be OK for you?

Would Tuesday at 9.0 suit you?

Could you manage next Monday?

**Checking if you are free**

Let me just check in my appointment book / diary.

I'll just have to see if I have any appointments that day.

Hang on, let me just check if I haven't got anything else on.

**Responding positively**

That's sounds fine.

That would be great. Nine o'clock next Wednesday?

**Suggesting alternative**

Actually Monday's a public holiday, is there any chance of having it on the Tuesday?

Could we make it a bit earlier / later?

Could we fix an alternative?

I've got a meeting all afternoon, how about the morning?

**Responding negatively**

Sorry, but I won't be available then.

Sorry but I'm tied up on Tuesday.

Sorry, but I'm going to be tied up all day.

**Confirming**

OK, three o'clock at your office.

Just so that we're clear. 10 a.m. on Tuesday the fifth?

**Changing / Canceling meeting**

I'm afraid I've got to call off tomorrow's meeting.

The thing is that I have to …

I hope you understand.

I hope I haven't messed up your arrangements too much.

I'm really sorry about this, but I'm afraid there's nothing I can do about it.

## 18.22 Arranging a meeting for a colleague

### Arranging the time

Good morning Mrs Jones, this is *name of company* calling. I'm Mr X's secretary.

I would like to arrange a meeting between Mr X and Mr Y.

Mr X will be in London on business from the 5th to the 8th.

Would it be convenient for Mr Y to meet up with Mr X on the morning of Wednesday the 6th in his office?

I'll send you a fax of Mr X's travel schedule.

Could you confirm by return whether the 6th is convenient and at what time would be most appropriate for their meeting?

If the 6th is inconvenient please could you let me know when Mr Y will be free.

### After confirmation

Is the office far from the airport?

How far is the office from the hotel?

Mr Rossi will be arriving on Wednesday the 6th of March.

Flight number AZ 4236 from Pisa.

He will be carrying a *name of company* brochure for identification.

He will be staying at the Palace Hotel.

The phone number is 0208 638 0041.

Could you arrange for Mr Rossi to be met at the airport?

Would he be best to take a taxi or can he just get the tube?

Where's the nearest station to your office?

Mr Rossi will be waiting in the hotel lobby at 10 o'clock.

### When the foreign company arranges a meeting at your office

Good afternoon, this is Mr Rossi's secretary.

If you could confirm all Mr Smithson's travel plans by fax.

I'll check Mr R's appointments and confirm when it would be convenient for Mr Rossi to meet Mr S.

I'm sorry but Mr R has an extremely tight schedule all that week.

Could we make it the week after instead?

I'm afraid that Mr R will be tied up all that afternoon, what about the following morning?

**After the arrangements**

Hello, could I speak to Mr S's secretary?

Just phoning to say that Mr R will personally pick up Mr S from the airport in Pisa.

Does he need me to make hotel reservations for him?

The best thing would be if Mr S could get a taxi from the airport,

We're only a ten minute drive away.

**Changing arrangements**

I'm awfully sorry, but I'm afraid Mr R won't be able to make the meeting on the 6th.

Could we possibly change it and make it the 7th instead?

If not, I'm afraid we'll have to postpone it till next month.

Do you think you could check with Mr S and ring me back?

I hope this doesn't mess up Mr S's schedule.

I'm afraid there seems to have been a bit of a mix up.

Thanks very much for your help - and please send Mr R's apologies to Mr S.

## 18.23   Hotel reservations

**Making the reservation**

This is *name of company* calling.

I would like to reserve a single room with bathroom …

   from the 8th to the 12th of November.

   for four nights in the name of Rossi.

He will be in on the 8th and out on the 12th.

Shall I give you the spelling of Rossi?

That's Rossi - R-O-S-S-I.

And our address is Via San Martino number 54.

That's V-I-A new word SAN new word MARTINO, number 54.

new line, 56125 Pisa, Italy.

He will be arriving late in the evening.

He should get to the hotel at about 11 o'clock.

How much is a single room with shower?

Is breakfast included?

Is there a choice between a continental and an English breakfast?

Mr R would prefer a continental breakfast.

Mr R will be paying with his American Express Card number 347-612-300-09 and it expires in *month year*.

Mr R will send you a sterling bank cheque to cover the advance payment of £ 120.

Please could you email confirmation.

By the way - who have I been talking to?

And your name is?

OK, goodbye Mrs Jones and thank you very much for your help.

**Changing and checking reservations**

You should have a reservation in the name of Rossi for tomorrow night.

Mr R will be checking in rather late, I'm afraid, at about 11 pm, you will hold the room for him, won't you?

I'm sorry but it looks as if I'll have to change Mr R's booking.


## 18.24  Saying goodbye

OK / Right, I think that's all.

Well, I think that's everything. Goodbye.

I look forward to seeing you / our meeting.

Do call if you need anything else.

Have a nice day / weekend.

You too.

Hear from you soon. Bye.

# THE AUTHOR

**Adrian Wallwork**

I am the author of over 30 books aimed at helping non-native English speakers to communicate more effectively in English. I have published 13 books with Springer Science and Business Media (the publisher of this book), three Business English coursebooks with Oxford University Press, and also other books for Cambridge University Press, Scholastic, and the BBC.

I teach Business English at several IT companies in Pisa (Italy). I also teach PhD students from around the world how to write and present their work in English. My company, English for Academics, also offers an editing service.

**Contacts and Editing Service**

Contact me at: adrian.wallwork@gmail.com

Link up with me at:

www.linkedin.com/pub/dir/Adrian/Wallwork

Learn more about my services at:

e4ac.com

# Index

This index is by section number, not by page number. Numbers in bold refer to whole chapters. Numbers not in bold refer to sections within a chapter.

**A**
Audio
  books, 15.10
  conference calls, **13**, 17.16
Authoritative and competent tone, 9.6

**C**
Calling another company for commercial reasons, 6, 17.12
Chasing, **7**
Complaints, 8.3
Conference calls, see Audio

**D**
Difficult callers, **8**
Dragon's Den, 15.4

**E**
Email, 1.2, 2.3
English level, 2.4, 5.6, **14**

**F**
First impressions, 1.5
Friendly relationships, 5.5, 5.9, 9, **11**

**H**
Helpdesk, **10**, **11**, **12**

**I**
Initial salutations, 2.1, 5.1
Instant Messaging, 13.7, 14.7
Introducing yourself, 2.1, 5.1

**L**
Listening, **14**, **15**, 16.1

**M**
Making calls, **2**, **6**
Meetings, 17.9, 17.10
Messages
  leaving, **3**, 4.2
  taking, 5.10
Movies, 15.7

**N**
Nervousness, 1.3
News as a listening source, 15.2
Note taking, 2.3

**O**
Orders, 7.2

**P**
Payments, 7.1, 17.11
Performance
  evaluation of, 9.7
Phrases, **18**
Podcasts, 15.9
Preparation, **1**
Present perfect, 5.7
Pronunciation, 10.3, **16**

**R**
Receiving calls, **5**, 6.1

**S**
Skype, 13.7
Songs, 15.9
Spelling names and addresses, 3.4–3.6
Subtitles, 15.8
Switchboard, **3**, **5**, 17.1–17.8
Syllable stress, **16**

**T**
TED, 15.6
Transcribing telephone calls, 1.4
Transferring calls, 5.2–5.4
TV series, 15.5
Typical phrases, **18**, see also Phrases

**U**
Understanding, **14**
Unhelpful staff, **8**
Useful phrases, **18**, see also Phrases

**V**
Video conference calls, **13**
Voicemail, **4**

**W**
Wrong numbers, 5.11

**Y**
YouTube, 15.3

Printed by Amazon Italia Logistica S.r.l.
Torrazza Piemonte (TO), Italy

45047168R00112